CW00724888

Plastics in the
automotive industry

Plastics
in the
automotive industry

JAMES MAXWELL

WOODHEAD PUBLISHING LIMITED

CAMBRIDGE, ENGLAND

Published by Woodhead Publishing Limited, Abington Hall, Abington, Cambridge CB1 6AH, England

Published in North America by the Society of Automotive Engineers, Inc, 400 Commonwealth Drive, Warrendale, PA 15096-0001, USA

First published 1994, Woodhead Publishing Ltd and Society of Automotive Engineers, Inc

© 1994, Woodhead Publishing Ltd

Conditions of sale

All rights reserved. No part of this publication may be reproduced or transmitted in any form or by any means, electronic or mechanical, including photocopy, recording, or any information storage and retrieval system, without permission in writing from the publisher.

British Library Cataloguing in Publication Data
A catalogue record for this book is available from the British Library.

Library of Congress Cataloging-in-Publication Data
A catalogue record for this book is available from the Library of Congress.

Woodhead Publishing ISBN 1 85573 039 1
Society of Automotive Engineers ISBN 1 56091 527 7

SAE order number: R-147

Designed by Geoff Green (text) and Chris Feely (jacket).
Typeset by Best-set Typesetter Ltd, Hong Kong.
Printed by Galliard (Printers) Ltd, Great Yarmouth, Norfolk, England.

The Plastic Age

Like us on a graph it will break or burn if it strays
Into the frost or flame where the axis looms.
Like us it's no better or worse than the endless ways
Its maker makes in his beautiful living rooms.
Glyn Maxwell

Contents

Preface

This book is not only a survey of the role and reach of plastics in automobiles, but is an attempt to build a bridge between two different but interdependent industries. It describes the application of plastics to each sector of automotive engineering, and identifies the factors involved in their selection. It reviews key historical developments and assesses future opportunities and constraints.

Primarily, the book is aimed at automotive engineers, designers and specifiers. Many are working with plastics materials and designing plastic components, having had no formal training in polymers, and using data sources which can be confusing and unhelpful. Experiences lecturing to graduate engineers have taught me that mechanical engineering and polymer science are cultures which are largely alien to each other. In spite of some brave attempts to combat over-specialization, the problem is still very widespread, particularly (but by no means exclusively) in Great Britain. In deference to engineers, therefore, no chemical formulae appear in this book. No specialized knowledge of plastics is assumed; merely a willingness to understand, and to work with (rather than against!) the essential physics and chemistry of these materials.

Secondly, the book attempts to steer polymer people towards an understanding of the needs and problems of automotive engineers. Awareness has been growing among polymer chemists and physicists in recent years, but in the past the plastics industry has shown an extraordinary reluctance to relate to the pressures of the assembly line, or even to present data in a relevant way.

The book consists of nine chapters. The first three constitute a comprehensive 'user's guide' to plastics, covering materials and processes generally, but with an automotive focus. The first chapter includes an historical summary of automotive plastics, and examines the problems as well as the benefits of the materials. The second chapter ('Understanding plastics') includes sections on composites, processes and functional design, and the third ('Choosing plastics') offers routes to material selection in the individual application sectors. The next four chapters address these application areas systematically, under the headings, respectively, of 'Interiors', 'Exteriors', 'Engine, power train and chassis' and 'Electrics'. The two final chapters examine the global problems generating much comment and concern at the time of writing; Chapter 8 is devoted entirely to that most pressing (but not necessarily most serious) problem – recycling.

Because of the high profile of global environmental issues, this book should interest anyone concerned about the future of the motor industry and the plastics industry. Already both industries are being targeted by well-intentioned but ill-informed legislation. It is important that policy-makers (in the boardroom as well as the legislature) are aware of the realities: that deep-seated changes, like designing and proving new cars or new recovery processes, take a long time; and that, if the energy factors are properly assessed, the benefits that plastics can provide far exceed the disadvantages.

Acknowledgements

I would like to record my appreciation of the help received from many sources. Generally, for their opinion-forming input received over many years, I thank my colleagues from the former ICI Plastics Division, whose concentrated polymer wisdom and experience has now been scattered over many fields, perhaps sadly, but certainly fruitfully. Specifically, for helpful discussions and the provision of photographs, I thank Albert Adams (Lotus Engineering), Tony Bardsley (Du Pont), Jan Czerski and Alfred Pirker (BASF), Kevin Dunn (ICI Polyurethanes) and Leo Morelli (Communication Systems, Brussels), Mehmet Sonmez (ICI Materials), Dick Thomas (ICI Acrylics), Laurent Rotival (Dow Plastics), John Whittaker (The Bird Group), Chris Wakley (Chris Wakley PR, for GE Plastics), S van Buijtenen and E Baeten (DSM Polymers) and D Westaway (DSM Resins).

Most importantly, I thank my wife for her sustained forbearance and support.

1

Materials for cars

Plastics usage

Materials and their conversion into components account for more than half the cost of a car. The materials mix is changing continually, and the fastest growing category is plastics. In 1993 a typical car contained some 100 kg of plastics, constituting around 11% of its total weight. Globally, this represents a market of over 4 million tonnes of plastics, with a value of over £5 billion.

Plastics usage in cars has been rising steadily since the 1950s, as Fig. 1.1 shows. When expressed in terms of percentage of car weight (see Fig. 1.2) the rate of growth appears to be increasing more dramatically; this is deceptive because over most of this period cars were becoming lighter. Average car weights fell by more than 15% between 1973 (when the cost of oil first became a major factor in automotive design) and the end of the 1980s. The highest growth rate for automotive plastics was in this period. It became evident that any new problems arising from the supply of oil-based plastics could be more than adequately compensated by fuel economies from plastics-based design changes and weight-saving techniques. Furthermore, greater cost consciousness highlighted the potential advantages of plastics in component consolidation and simplified assembly.

Some extravagant predictions have been made in the past for the un-stoppable growth of automotive plastics. There are clear signs now however that the growth curve is beginning to flatten out. An ultimate levelling out is inevitable in any case, because of the irreplaceable metal parts. Excessive weight reduction would lead to unacceptable handling characteristics; how-ever, the current trend towards stronger body structures is once more increasing the average weight of cars. Increasing competition and prolonged recession are encouraging a cool appraisal of the cost and performance of all materials. In this chapter both the benefits and the limitations of plastics are examined in a general context, with a look at the historical background and at the very significant changes in the types and qualities of the plastics which the motor industry selects. Table 1.1 summarizes the classes of materials to

1

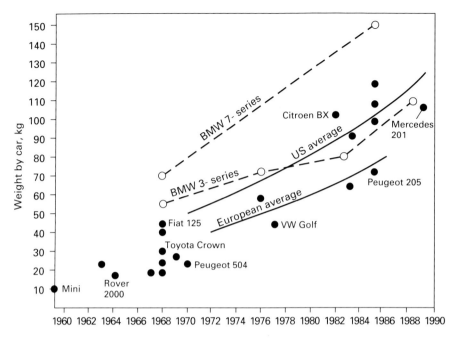

1.1 Growth of plastics content in cars, by weight.

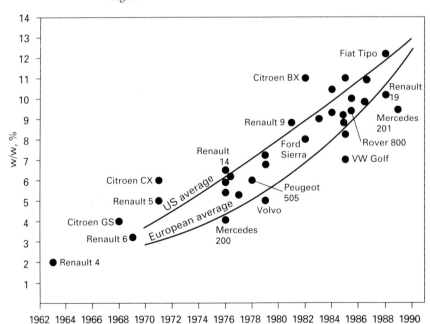

1.2 Growth of plastics content in cars, by weight percentage.

Table 1.1. Material content of a typical car by weight (%)

Steel sheet	55
Cast iron	10
Plastics	10
Non-ferrous metals	7
Liquids	5
Elastomers	4
Glass	3
Textiles and cellulose products	2
Insulating materials	2
Paints and adhesives	2
	100

Table 1.2. Typical figures for plastics consumption by type (%)

Polypropylene (PP)	23
Polyurethane (PUR)	21
Polyvinyl chloride (PVC)	12
Acrylonitrile-butadiene-styrene (ABS)	10
Thermosetting polymers	8
Polyamides (PA, nylon)	8
Polyethylene (high and low density, HDPE and LDPE)	5
Polycarbonate (PC) and blends	4
Polyphenylene ether/oxide (PPE/PPO) and blends	3
Polyformaldehyde (POM, acetal)	2
Polymethyl methacrylate (PMMA)	2
Thermoplastic polyesters (PET and PBT)	1
Others	1
	100

be found in a modern car; Table 1.2 provides a similar summary of the types of plastics used by the motor industry. These are very general figures; there are considerable differences between models, and regional differences between Europe, America and Japan.

History of automotive plastics

The motor industry was using plastics long before the Second World War, in such items as electrical components and interior fittings. They were used whenever they were available as a convenient low cost alternative to traditional materials. These were direct replacement applications; the concept of redesigning around the essential merits of plastics came much later, following some fundamental changes on the world scene.

The first major change was the development of low cost oil in the post-war years, providing a reliable and consistent raw material for cheap plastics.

Prices of bulk plastics fell, and continued to fall during the 1950s and most of the 1960s. The second big factor was the appearance, in this same period, of a seemingly endless flow of exciting new materials. Most of the materials now classed as engineering plastics were in volume production in their basic formulations before the end of the 1960s. The spectacular growth in the plastics market since the early 1970s has resulted not so much from the discovery of new plastics as from the refining and targeting of the established ones. In Fig. 1.3 the major discoveries in plastics are plotted against a background of the growth in the total world plastics market.

From the beginning, much of the development work on engineering plastics was directed towards the motor industry. These materials (together with reinforced composite versions of some bulk plastics) gradually acquired respectability in the automotive industry, as it came to be seen that they could be produced to consistent formulations, and held to precise engineering specifications. The development of a proper understanding between the automotive and plastics industries did not happen overnight; in fact it is an ongoing process, and there is still room for improvement. Nevertheless, it is

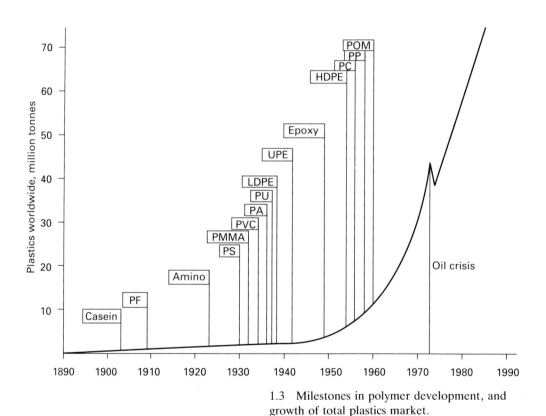

1.3   Milestones in polymer development, and growth of total plastics market.

generally true today that plastics are used not reactively but proactively. The new plastics applications are there because of factors like improved safety, easier assembly, cheaper maintenance, new design possibilities and better cost/performance relationships.

Table 1.3 is a calendar of the development of plastics applications in the

Table 1.3. Plastics milestones in the automotive industry

Date	Innovation in plastics	Material	Model/source
Pre–1950	Distributor, switches, etc	PF	
	Steering wheel shroud	PF	
	Coloured knobs and handles	CA, CAB	
	Leathercloth	PVC	
1950–1954	Cable sheathing	PVC	
	Car body (panels)	GRP	Chevrolet Corvette
	Battery case (in car)	PS	NSU Prinz
	Gear wheels (wipers, speedo, etc)	PA	
	Rear light lenses	PMMA	
1955–1959	Roof	GRP	Citroen DS
	Truck cabs	GRP	
	Rotationally cast arm rests etc	PVC	
	Vacuum formed interior panels	ABS	
	Foam-backed fascia skin	PVC	Mercedes
	Trim fasteners	PA	
1960–1964	Integral hinge: gas pedal etc	PP	GM USA
	Injection moulded cooling fan	PP, PA	Rootes, BL
	Improved distributor cap	GR Alkyd	
	Carpet backing	LDPE	
	Door and window components	PA, POM	
	Rigid fascia covers	PP, ABS	
	Kick plates	PP	Vauxhall
1965–1969	Injection moulded heater case	PP	
	Seat foam	PU	
	Front grille	ABS	Renault 6
	Expansion tank	PP	
	Hydraulic fluid reservoirs	PP	
	Headlamp lenses	PC	USA
1970–1974	Battery box (engine compartment)	PP	Lucas
	Solderless radiator end tanks	GR-PA	Sofica, for VW
	Bumpers (rigid)	SMC	Renault 5
	Lateral protection panels	SMC	Renault 5
	Fuel tank	HDPE	VW Passat
	Front and rear spoilers	PU	BMW
	Two-piece car body	GRP	Lotus Elite
	Headlamp body	DMC	Lucas, for Vauxhall
1975–1979	Painted RTM panels	Epoxy	Matra Murena
	Door-mounted mirrors	GR-PA	Renault 14
	Painted grille	SMC	Ford Cortina
	Flexible lateral protection strip	PU	Mercedes
	Flexible bumper covers	PP/EPDM	Fiat, Citroen
	Engine compartment under-panel	GMT (PP)	USA
	Oil pan	STX	GM USA
	Estate car seat-back/load floor	HDPE/GMT	USA

Table 1.3.   *(Continued)*

Date	Innovation in plastics	Material	Model/source
1980–1984	Bonnet and tailgate (volume prod.)	SMC/XMC	Citroen BX
	Self-supporting bumper	PC/PBT	Ford Sierra
	Painted thermoplastic panel	ABS, GR-PA	Ford Sierra
	Full-face wheel trim	GR-PA	Ford Sierra
	Slush-moulded fascia	PVC	Audi Quattro
	Bumper beam	SMC	Ford Australia
	Car body (panels, volume prod.)	Various	Pontiac Fiero
	Painted PP bumpers	PP	Rover 200
	Painted thermoplastic body panels	PC/ABS	Honda CR-X
	Steering wheel + integral skin foam	PU	USA
1985–	On-line-painted TP body panels	PPE/PA	Nissan Be-1
	Painted RTM panels (volume prod.)	Epoxy	Renault Espace
	Inlet manifolds (fusible core)	BMC/GR-PA	Ford, BMW
	Body platform	GR-PU/Epoxy	BMW Z1
	Injection moulded under panel	GR-PP	Volvo 400
	Composite drive shaft	CR + CR Epoxy	Renault Espace (Quadra)

motor industry. The indicated dates refer to series production, large or small; prototypes are excluded. As the events cannot all be dated precisely, they are grouped in five-year periods. (Inevitably, this is a subjective list which cannot claim to be comprehensive.)

Why plastics?

No longer considered simply as replacement materials, plastics are now employed for their merits, for the new effects they can provide, and the cost benefits which follow. A consensus view from the motor industry of the benefits of plastics might list the following factors in order of importance:

- Economy.
- Weight reduction.
- Styling potential.
- Functional design.
- New effects.
- Reduced maintenance.
- Corrosion resistance.

Typical examples are quoted here in support of each of these points.

Economy

The 'macro' price factors tend to favour plastics in comparison with metals. The energy involved in production and conversion is lower (particularly so in relation to aluminium), and the prices of plastics have tended to remain remarkably stable in times of fluctuating oil prices. Bulk polymers apart, direct costs in terms of price per kilogram are less favourable to plastics. Translating into volume prices usually redresses the balance, and, in truth, in relation to components whose dimensions are predetermined, volume price is a more realistic measure than the conventional weight price.

Raw material price, however, is only the beginning of the story. From an automotive designer's viewpoint, the critical parameter is the 'piece part' cost. This, of course, is derived from several factors:

- Raw material cost.
- Tooling cost.
- Direct conversion cost per part.
- Total number of parts.
- Cost of sub-assembly.
- Assembly line costs.

The last three factors can be transformed by what is probably the biggest single benefit of plastics: the potential they offer for part consolidation. This is where an assembly consisting of several discrete components is replaced by a single part made in a single processing operation.

In approaching the question of converting a component to plastics, the prime consideration usually involves 'The numbers game': i.e., do the planned production numbers justify the up-front tooling costs? The answer will determine the choice not only of the material, but also of the conversion method.

Weight reduction

Weight saving is not necessarily of direct interest to the motor manufacturer. However, fuel economy is of interest to customers and car manfacturers alike, and in the USA severe cost penalties were imposed on the motor manufacturers after the 1973 oil crisis, as one means of trimming the enormous fuel consumption in the USA. It has been calculated that automotive weight reductions directly attributable to plastics lead to a 5% saving of fuel. Globally, this could amount to a saving of over 15 million tonnes per year.

In some metal replacement applications the weight saving is spectacular. A fuel tank in high density polyethylene (HDPE) represents a saving of around 40%; in battery boxes the figure can be as much as 70%. Frequently redesign in plastics will still achieve a weight saving even though wall thicknesses have been increased to maintain rigidity.

Styling potential

The benefits of plastics in terms of styling are not easily quantified. The fact is that plastics have changed almost every aspect of car appearance. In good design, of course, styling and function are integrated. The modern adjustable door-mounted mirror, aerodynamically styled and sometimes colour matched, is a much more practicable item than its predecessor: exterior lower side panels, be they in rigid SMC (sheet moulding compound) or flexible polyurethane (PUR), are strictly functional items which nevertheless make a major contribution to the appearance of a vehicle.

The evolution of traditional car bumpers into the sophisticated front and rear ends of today has been facilitated at every step by the versatility and almost unlimited shaping potential of plastics. Styling has had to be compatible with a daunting list of attributes like energy absorption, creep resistance, dimensional stability, chemical and abrasion resistance, temperature and UV performance and paintability. These requirements could possibly have been met by other materials, but styling would undoubtedly have suffered.

Functional design

What plastics can do for design goes far beyond styling. Probably the most important single attribute conferred by plastics is design freedom, in terms of changing shapes to fit the space available, and (most important of all) the possibilities for component consolidation. The fuel tank (see Fig. 6.24) is an excellent example of the former; modern blow moulding techniques allow for the design of highly asymmetrical forms with severe re-entrant angles, without losing the ability to include inserts and control wall thickness. A good example of consolidation is the underbonnet front end support, as employed in the Peugeot 205 (Fig. 1.4). An assembly involving more than twenty steel shaping operations is replaced by a single plastic part. Injection moulded glass reinforced nylon and polypropylene, glass mat thermoplastic (GMT) polypropylene, and epoxy-based SMC are all used in different Peugeot models for this application. The choice depends on the strength and rigidity requirements, the wastage inherent in the design, and the length of model run.

New effects

The 'new effects' category of benefits includes a wide range of characteristics such as sound dampening, thermal and electrical insulation, energy absorption, and the host of possibilities for marrying different characteristics, for

1.4   Consolidation in plastics: front panel from Peugeot 205 (courtesy ICI).

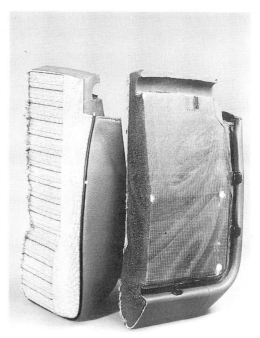

1.5   Section through Renault 5 seats, polyurethane foam produced *in situ* (courtesy ICI Polyurethanes).

example, foam backing of flexible surfaces or fabric, and integral skinning of foams. Figure 1.5 shows sections of seats in which the foam has been produced in direct contact with the fabric. A great number of effects such as these in fact result from the availability of well-engineered plastic foams, primarily polyurethane.

Implicit in all this is the ease with which different useful properties can be found in the one material, and the way in which surface appearance can be changed at will in terms of colour, gloss and texture. The safety standards of the modern vehicle, with all the improvements designed for passenger and pedestrian protection, derive from these characteristics of plastics. So does the revolution in the level of comfort and passenger-friendliness of car interiors which the past forty years has seen.

Reduced maintenance

The much greater reliability of vehicles and their more relaxed maintenance schedules are largely due to the plastics revolution of the past forty years. Essentially it is the characteristics of lubricity and load spreading which are at the heart of this revolution. The materials most concerned are those engineering plastics that have a favourable balance of rigidity and resilience, with useful resistance to abrasion, temperature, fatigue and relevant chemicals. They span a price range from the familiar nylons and acetals to inert, high temperature specialist polymers with a higher temperature capability.

The automotive applications of these materials cover a huge variety of moving parts; gears, cams, bearings, ratchets, sliders and valve seats are just a few examples. Beyond reduced maintenance, there are subsidiary benefits in many cases like vibration dampening, power reduction and quietness.

This is now a specialized area of product development; a wide range of formulations has been derived from the small family of basic engineering plastics. The principal additives are the reinforcing fibres (glass, carbon, aramid, etc.), and lubricants such as PTFE, silicone, graphite and molybdenum disulphide, each of which makes a specific contribution.

Corrosion resistance

Each polymer has its own distinctive chemical nature. Some are affected by fuels and lubricants, others are attacked by battery acid, and all are visually affected to some degree by weathering. However, none is susceptible to factors like acid rain, sea spray and road salt which can do so much damage to unprotected sheet steel, requiring the automotive industry to spend huge sums on corrosion protection.

Problems with plastics

Bridging the confidence gap

Designers contemplating a plastics solution to an application problem often feel a lack of confidence. We have all been exposed to the poor image of plastics; not surprisingly, experiences of failures in 'low tech' applications can deter designers from adventures in 'high tech' areas. To bridge this confidence gap it is important to appreciate that there are basic behavioural differences between metals in general and plastics in general.

Most aspects of the behaviour of materials derive from their chemical nature. However the aim here is to improve understanding without recourse to the detailed nomenclature and formulae of polymer chemistry. Nevertheless, this does not relieve the reader of the need to 'think molecules'. Polymers on the 'micro' scale are not wholly homogeneous materials, and to understand their behaviour it is necessary to visualize their long chain molecules.

Every point of difference between metals and plastics is a potential trap for the unwary. Nevertheless, the performance of plastics is predictable and consistent. The essence of a successful application is to avoid serious mistakes in respect of design, material choice and production. Every little disaster with a plastic component can be traced to one of the three prime factors detailed here.

Performance problems

The actual performance problems which result from mistakes made in design, material choice and production usually manifest themselves as failures of dimensions, creep, brittleness or chemistry. Dimensional control in particular needs careful understanding. There are the familiar problems of metals, plus some additional ones.

Failure by deflection can occur when exposure to temperature or to sustained loading induces excessive molecular movement. This is the phenomenon of creep.

Failure by fracture occurs when there is insufficient molecular movement to accommodate an imposed load. Very often, when a material fails to give its expected impact performance, either bad design or bad processing is to blame. Chapter 2 contains ideas on how to avoid brittle fracture.

There remain the performance problems which can be attributed to chemistry, such as the direct problems of chemical attack, staining, absorption and plasticizer loss. There are the very specific problems of adhesion to paints and adhesives, and there are the problems of surface ageing, induced by heat, ultra-violet light and radiation. Finally there is the very complex question of burning behaviour. These factors are all part of the plastics equation, along with the numerous advantages. Chapter 2 ('Understanding plastics') and Chapter 3 ('Choosing plastics') attempt to demonstrate how to maximize the benefits and minimize the disadvantages.

Compatibility with automotive industry practices

When plastics progress beyond separate components or minor parts of sub-assemblies, they run into a different order of problem; the problem of being incompatible with existing motor industry practices. This concerns not only the birth of the vehicle, but also its demise; not only its construction, but also the stripping and shredding processes.

It is clear that plastic body panels are effectively incompatible with existing body shop practices. Although plastic sheet raw materials are used widely and successfully, their stamping and shaping processes are quite different, and much slower than those of the press shop. Furthermore plastics are not readily spot welded to metals or to each other. Electrostatic painting is not applicable to plastics, and not many polymers can stand up to topcoat oven temperatures. Even introducing plastic panels in a small way involves significant changes to body shop procedure, while a wholly plastic construction necessitates completely new concepts. The key factor is the scale of production. The problems involved with the different materials and processes, and the opportunities they offer, are discussed in Chapter 5.

In the last few years, the question of the disposal and recycling of non-metallic automotive materials has assumed great importance. In essence, the problem centres on the fact that the plastics content of vehicles, hitherto discarded and expended as landfill, has increased to the extent that it threatens the viability of the whole operation. Designing for recycling is now a very active field of study. In the short term, however, widespread concern about the possibility of new legislation is dampening the enthusiasm for plastic body panels (see Chapter 8).

2

Understanding plastics

Classifying plastics

The term 'plastics' covers a broader variety of species than the term 'metals'. Hence, without effective classification, the possibilities for over-simplification and confusion are endless. In terms of mechanical performance in automotive applications, and in particular the response to stress, probably the most important distinction is between unfilled plastics and composites. In automotive plastics terms, 'composite' means plastics reinforced with fibres, usually glass. Nowadays there is a very extensive range of distinct materials answering to this description; their significance is discussed later.

Perhaps the most confusing aspect of plastics to the non-specialist is the profusion of raw material forms and processes involved. These are classified and reviewed in this chapter, together with the all-important questions of economics and 'The numbers game'. Polymers can themselves be classified in a number of ways:

- Thermoplastics and thermosets.
- Chemical structure.
- Crystalline and amorphous thermoplastics.

Thermoplastics and thermosets

Essentially, polymers can be defined as either thermoplastic or thermo-setting. Although they often display similar properties, and indeed in the automotive industry frequently compete for the same applications, there are fundamental differences in structure and processing methods. These differ-ences have implications for production costs and feasibility.

Thermoplastics are softened and melted by heat, and shaped or formed before being allowed to freeze again. The heating and freezing processes can be repeated many times without significant chemical change (although not without some deterioration, usually attributable to the breaking of molecular chains).

13

Thermosetting materials, when heated above a critical temperature, undergo chemical reaction involving cross-linking between chains. Shaping or fabrication must be effected before this cross-linking is complete, because the thermosetting process is not reversible. Thermoplastics in general exhibit better flexural and impact performance and superior resistance to key solvents; thermosets tend to have better compressive strength and abrasion resistance and significantly better dimensional stability.

The response to temperature reflects the difference in molecular structure. Thermosets are made up of giant cross-linked molecules which undergo very little movement under thermal or mechanical stress. At high temperatures they will lose some volatile material and eventually char, but they will not melt. In thermoplastics, the separate molecular chains become increasingly mobile as the temperature rises, until the material softens or melts. With any further rise the melt becomes increasingly fluid, and eventually thermal decomposition ensues.

Most of the high-volume automotive polymers are thermoplastics; the common virtues of polypropylene, ABS (acrylonitrile butadiene styrene), PVC (polyvinyl chloride), nylon, polyethylene and polycarbonate etc., are their versatility, ability to be injection moulded or extruded into intricate shapes, and suitability for mass production. Thermosets used in significant volume include phenolics, long used as electrical components, and now finding application in non-burning creep-resistant underbonnet items, and glass-reinforced polyesters, used originally in hand lay-up processes, but nowadays widely used in SMC.

Plastics and elastomers

This classification is based on differences in mechanical performance; specifically, differences in the strain response to applied stress. 'Elastic' behaviour is defined by the deformation being proportional to the stress, and fully recoverable when the stress is removed. True 'plastic' behaviour implies continuing flow under load, without recovery afterwards. In practice, polymers are 'viscoelastic', their behaviour being partly plastic and partly elastic, with the balance between the two extremes varying with the temperature and time under load.

Elastomers are not normally regarded as plastics, but the distinction is not always helpful, because the two categories have many characteristics in common and in any case frequently co-exist in the same sub-assembly. What the designer needs to know is that there is now available a wide variety of materials described as elastomers, which are readily deformed by low stresses, and revert rapidly and (almost) completely to their initial dimensions when the stress is removed. These are the attributes of a rubber.

Natural rubber is itself a thermosetting elastomer. There are cross-linkages between the molecules, preventing 'sliding' between molecular chains and ensuring recovery from load, but normally these are few enough and distant enough to preserve an overall flexibility. Rubbers can be formulated to give a much higher number of cross-links, making the material hard and rigid. Synthetic thermosetting elastomers can likewise be formulated with different degrees of cross-linking, to change their property balance.

In recent years, several types of thermoplastic elastomers have been developed. There are still cross-linkages, but they are weak and flexible compared with thermosetting elastomers. Elasticity arises from the chains themselves being a mixture of rigid and flexible regions. Thermoplastic elastomers are also very versatile in formulation: here the differences are achieved by changing the balance between 'hard' (crystalline) and 'soft' (amorphous) parts of the chains, rather than by changing the degree of cross-linking.

The new elastomers are particularly relevant to the automotive industry because they offer better properties – particularly heat, oil and fuel resistance – than the established materials such as natural and synthetic rubber and plasticized PVC. Among the most important types are PUR elastomers, PBT block copolymers, EPDM olefinic terpolymers and ethylene-acrylic elastomers. Typical applications are the traditional 'rubbery' ones of gaskets, seals, gaiters and cable covers, but set in the aggressive underbonnet environment of today's performance vehicles. Beyond this, however, there are examples where these materials are sufficiently versatile to have been selected, sometimes with reinforcement, as engineering components in their own right.

Chemical types

The chemical structure of polymers is all-important in determining the behaviour of the raw materials of plastics. Some useful guidelines follow.

Most thermoplastics are made up of molecules with backbones consisting of a linear chain of carbon-to-carbon links. This description fits bulk polymers like polyethylene and 'engineering plastics' like nylons; it also fits the fluoropolymer PTFE (polytetrafluorethylene), an extreme case, where the chains are so compact and so long that the material cannot be melted and must be processed by sintering.

Many of the more recently developed thermoplastics are made up mainly of linked ring systems. These are very stable structures, and more resistant to heat, fire and chemicals than straight chain polymers. Most are classified as 'aromatic', because structurally their molecules include benzene rings directly in the chain. Their individuality (and their nomenclature) derives from the chemical groupings which link the benzene rings; hence polyether sulphone (PES), polyphenylene sulphide (PPS), polyetherether ketone (PEEK), etc. Several of the well-known engineering plastics are semi-aromatic polymers, with enough ring systems in the chain to be intermediate in terms of thermal, chemical and burning performance. Examples are polycarbonates and the thermoplastic polyesters polybutylene terephthalate (PBT) and polyethylene terephthalate (PET).

With thermosets there is an added order of complexity, through cross-linking. Increasing cure results in increased cross-linking, so that any one material can manifest a range of properties depending on the degree of cure. Increased cure improves the performance of engineering materials, but it makes them more difficult to process. A higher aromatic content can also improve the performance of thermosets. For example, epoxy vinyl ester resins generally display higher thermal and mechanical performance than the traditional unsaturated polyesters, because there are more aromatic groupings in the chain.

Crystalline and amorphous thermoplastics

The degree of crystallinity of a thermoplastic can be a useful guide to its behaviour. Crystallinity arises when the polymer chains are regularly oriented; normally this is only achieved intermittently throughout the mass of a material. Polymers in which the molecules are prevented from alignment by bulky side chains are amorphous. These differences in sub-microscopic

Table 2.1.   Crystalline and amorphous plastics in service: examples from the automotive industry

Property	Application	
	Crystalline	Amorphous
Transparency		PMMA for rear light lenses; PC for headlamp lenses
Solvent resistance	PA (nylon) for oil filler caps, petrol tubing, etc; PP for hydraulic reservoirs	
Fatigue resistance	PA for bonnet catches; acetal for door striker plates, etc	
Dimensional stability		PC/ABS blends for cowl panels; PPO/PA blends for wing panels
Properties relatively stable with temperature		PES for very high temperature electrical connectors
Easy flow melt	PP for one-piece roof and pillar liners; PA for cable ties	

structure or morphology lead to differences in behaviour patterns. The automotive applications listed in Table 2.1 illustrate these differences.

Table 2.1 simply lists a few guideline examples: it must be emphasized that the situations are rarely 'black and white', and moreover the crystalline/ amorphous attributes are not always the crucial ones.

Composites

What composites achieve

From straw-filled bricks through wattle-and-daub buildings down to plywood and reinforced concrete, it is clear that composite structures have always been vital to man and his works. The advent of sophisticated polymers and fibres, and new techniques for combining them, has enormously extended the scope and capability of materials.

It can be said that composites combine many of the best features of both metals and plastics. It is a complex picture, and as always there is a price to be paid for any benefit. Table 2.2 attempts to provide an overview of the complete landscape for metals, plastics and composites, picking out the highlights of performance capability. It is a simplistic, qualitative assessment, with just two levels of performance: better or worse. What Table 2.2 tells us is that, for the most part, composites enjoy the best of both worlds, with many of the most useful features of both metals and plastics.

The middle section of Table 2.2 indicates the potential trouble areas for all kinds of fibre-reinforced materials. Strength can be very high indeed, but

Table 2.2. Why use composites?

Property	Metals	Composites	Plastics
Strength	○	○	●
Creep resistance	○	○	●
Low thermal expansion	○	○	●
High temperature rigidity	○	○	●
Low temperature impact	○	○	●
Crushability	○	●	●
Isotropy	○	●	○
Surface finish	○	●	○
Plastic deformation	●	●	○
Lightness	●	○	○
Fatigue	●	○	○
Lubricity	●	○	○
Corrosion	●	○	○

○ = better; ● = worse.

instead of the ductile failure of 'tough' plastics under ordinary conditions, or the 'crushability' of steel under extreme conditions, the failure mode is brittle. When the fibres are long and the contact between fibres and matrix is good, the failure takes the form of splintering or delamination, in which the specimen retains its integrity. Failure is therefore much less catastrophic than with metals not exhibiting the crushability of steel sheet.

Surface finish can be a problem with composites, although there are many well-known examples of painted composite surfaces which are judged to be of 'Class A' surface finish. However, the most consistently under-appreciated problem with fibre-reinforced composites is anisotropy, i.e. the directional dependence of mechanical and dimensional properties. It arises from the orientation of the fibres, and consequently is most pronounced when a component has been shaped by a high speed melt flow process; injection moulding is the prime example. It is best to assume that anisotropy is always present in a composite, unless isotropy has been designed into the material; either by the use of a random glass mat, or by a deliberate layering process in which the orientation in different layers is balanced out.

Varieties of automotive composites

The variety of materials and processes in use can appear quite bewildering. Figure 2.1 summarizes all the thermoplastic and thermoset materials and processes which are relevant to the motor industry. This omits some of the

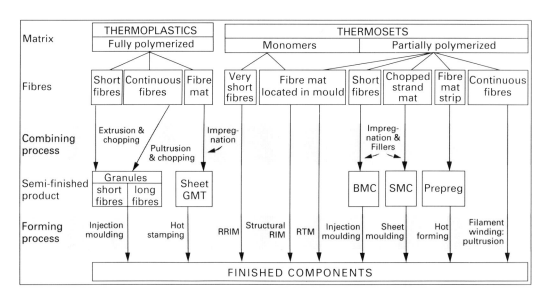

2.1   Summary of automotive composites.

more basic manual processes, together with some of the more speculative new processes emerging from the aerospace industry. Also omitted from this survey are fillers, the function of which is essentially to increase the stiffness and deformation resistance or to reduce the cost of polymers, and textile reinforcing agents. These (in phenolics in the 1920s) were in fact the original polymer composites.

The principal reinforcing fibre is of course glass, and specifically E-glass. The dominance of glass is unlikely to be challenged, at least in the motor industry. Carbon fibre is becoming increasingly important in aerospace applications, because of its greater strength and frictional performance. Apart from spectacular but rarified application in high strength structures for Formula One racing cars, its main automotive uses are in reinforced engineering thermoplastics for moving parts in engine and transmission. Aramid fibre is used as a reinforcing fibre in moving parts where lubricity and dimensional stability are more important than strength or rigidity. Asbestos was becoming popular as a low cost reinforcing fibre in PS, PP and PA until it was recognized as a health hazard in the 1970s. It has now disappeared from the product ranges of the raw material suppliers.

The properties of composites determine the processing techniques used, and these can be very different indeed. The two poles of the spectrum of composites are represented on the one hand by materials suitable for high speed conversion processes, such as injection-mouldable reinforced nylon, and on the other hand by materials used for high strength structures in very small numbers, like glass reinforced polyester for the hull of a minesweeper or carbon fibre reinforced epoxy for an airframe.

The pultrusion process involves drawing continuous longitudinal fibres through a resin impregnation bath, and then curing rapidly. This results in composites with exceedingly high strength and stiffness in the favoured direction. When a forming operation takes place during cure, the process becomes 'pulforming'. This is the basis of developments such as composite leaf springs and some safety steering columns.

Figure 2.1 is simplistic because it shows only the principal materials and processes. What is happening generally in development is a tendency for the outlines to become blurred. The reason is the desire to marry the two extreme benefits of high performance and high speed production. To this end, the advantages of one technique are continually being applied to another. There are many examples of this, such as:

● Thermoplastics are being incorporated in some thermoset prepregs to increase toughness. So far this is centred on high performance structural composites for aerospace.

● The application range of glass reinforced nylon is being extended by the

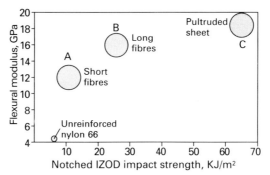

Comment: Results all relate to Nylon 66
reinforced to 50% w/w glass fibres.

Modulus figures all in most favoured direction:
A and B through fibre orientation in injection
moulding; C because samples cut in direction of
continuous fibre reinforcement.

2.2   Effect of fibre length on stiffness and
toughness of thermoplastic composites.

development of grades which have much longer fibres, but are still injection mouldable. The first of this type, an ICI material, is produced by pultrusion rather than by extrusion compounding. The fibres experience far less attrition before the lace is chopped into mouldable granules. As a result, these long fibre injection-mouldable grades, in terms of strength, rigidity and failure mode, achieve an intermediate status between short fibre moulding grades and long fibre composites produced by labour-intensive processes. This effect is shown schematically in Fig. 2.2.

● The idea of producing a finished component by locating the reinforcement where it is most needed, and then inducing the matrix to polymerize around it, was first used in hand lay-up processes, was taken up in resin transfer moulding and then in structural reaction injection moulding (S-RIM) systems. The same concept has now been extended to SMC, by replacing part of the random glass mat with uni-directional glass roving. The resulting 'high modulus continuous' SMC, or HMC, has very high strength and rigidity in the preferred direction, and is finding use for bumper beams.

It is recognized now that in terms of properties, the high performance composites have progressed about as far as can be reasonably expected. Likewise the problems (and the advantages!) of anisotropy can be very effectively addressed by computerized stress analysis techniques. Efforts are now being directed towards the fabrication processes, to make them faster, more mechanized and more consistent. SMC and BMC processes are already largely 'industrialized'; attention is being focused on such processes as RIM, RTM, filament winding and GMT stamping. In general, it is easier to control the precise location of the uni-directional reinforcement in 'in-mould' processes such as S-RIM and RTM, than in sheet-forming processes like GMT and SMC. Hence, higher strengths have been claimed for S-RIM in strongly directional applications such as bumper beams.

Processes

Introduction

Table 2.3 is a summary of the important processing techniques for plastics. Those of most relevance to the automotive industry are discussed in this section. Table 2.4 compares process capability in terms of (A) detail reproduction, (B) open or hollow, (C) Class A finish, (D) 1 or 2 surface forming.

It is clear that designing plastic components is impossible without a sound understanding of what is practicable for each class of material and each process. It is necessary to know what the constraints are in terms of size, shape, cost and final appearance.

Infusible polymers cannot be produced by these techniques. Hence, un-modified PTFE is frequently used as a sintered coating, or as an additive in special bearing compositions. Polyimide (PI) is even more difficult to process, and is normally supplied direct from the polymer manufacturer in the form of precision-sintered mouldings.

Table 2.3. Processes for plastics

Raw material form	Thermoplastics	Thermosets
Monomers and liquid resin	Casting	RIM*, R-RIM*, S-RIM* RTM* Hand lay-up, filament winding, fibre spray
Powder and paste	Dip coating, slush moulding*, calendering, fluidized bed coating, rotational moulding*, sintering	Compression moulding
Granules	Injection moulding*, extrusion*, blow moulding*	Injection moulding*
Sheet	Vacuum forming*, pressure forming*, GMT stamping*	Stamping of SMC* Forming of prepregs

*Particularly relevant to the automotive industry.

Table 2.4.   Capability of processes

Process	A	B	C	D
Injection moulding	Good	Open	Yes	Double
Structural foam moulding	Good	Open	No	Double
Blow moulding	Good	Hollow	No	Single
BMC moulding	Good	Open	Yes	Double
SMC moulding	Good	Open	Yes	Double
GMT stamping	Good	Open	No	Double
Rotational moulding	Limited	Hollow	No	Single
RIM	Good	Open	Yes	Double
RTM	Good	Open	Yes	Double
Thermoforming	Limited	Open	Yes	Single

Injection moulding

Injection moulding is the prime process for the manufacture of precision components from plastics in large numbers. Used predominantly for thermoplastics, it involves melting or plasticizing the granulated material in a heated barrel with a reciprocating screw, and then injecting it at high pressure into a closed mould. The operational sequence of plasticization, injection, cooling and ejection is controlled on a very precise time cycle. Modern injection moulding machines are massively constructed and extremely sophisticated; a moulding shop with a broad capability in terms of component size is necessarily an expensive investment.

The automotive designer will be concerned with the costs involved with a specific component. Moulding tools are likely to be made from hardened steel to extremely precise dimensions and involve highly skilled designers. There is therefore a considerable 'up-front' cost component in any injection moulding project. Expert calculations are necessary to relate the factors of annual production figures, model life, rate of call-off, etc., to the number of tools required, or (in the case of small components) the number of mould cavities per tool. Figure 2.3 shows multi-cavity mouldings, in the form in which they are extracted from the tool. A large intricate moulding like an instrument panel (see Fig. 2.4) presents a different kind of design problem. Organizing the tool layout to get the optimum flow pattern requires the use of all the best modern techniques of tool design. The overriding advantages which justify all this trouble and expense in so many cases are these:

- Design freedom, allowing wide variations in shape and form.
- Component consolidation, allowing drastic reduction of numbers of parts and assembly operations.
- Lower piece part costs in volume production.

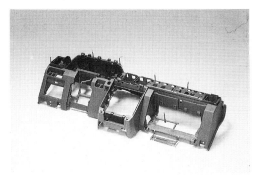

2.3 Nylon cable ties: an example of multi-cavity injection moulding (courtesy ICI).

2.4 Multiple gating of a complex injection moulding tool: instrument panel for Nissan Bluebird in polypropylene (courtesy ICI).

The nature of conventional injection moulding means that it is impossible to produce closed shapes, or ones with excessive undercuts. Extremely complex shapes can be injection moulded by incorporating sliding cores in the tool, but these will entail higher costs and longer cycle times. There is often a 'trade off' between higher tool costs and higher direct moulding costs.

The scope of the injection moulding process has been extended in recent years by some very ingenious elaborations:

– Multi-colour moulding. This concept was devised for the automotive industry, specifically for rear light lens clusters in PMMA (polymethyl methacrylate). It involves a rotary mould assembly, with the different colours being injected sequentially from separate cylinders as the mould rotates.
– Structural foam moulding. In its simplest form, this incorporates a chemical blowing agent in the moulding powder granules, resulting in a lightweight rigid foam structure with a very high stiffness to weight ratio. The limitations of the process are a poor surface finish and a long cycle time. The high pressure improvement of this method gives a solid skin with a good surface finish.
– Gas injection systems. The essence of these systems (associated primarily with Cinpres and with Battenfeld) is that the gas is localized, more or less down the middle of the section of the moulding, and not dispersed throughout the material. The high stiffness/weight ratio and good surface finish are retained, with the potential functional advantage of a central cavity.
– Sandwich moulding. This is a way of combining the best features of two materials by injecting them sequentially through the same sprue, so that

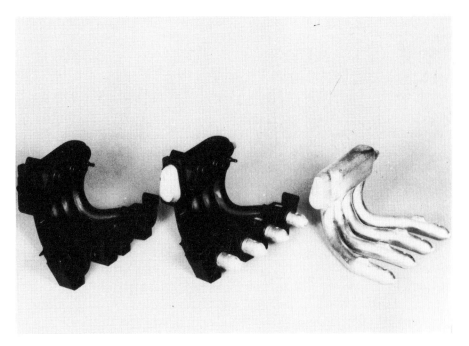

2.5   Fusible core injection moulding of an
intake manifold in glass reinforced nylon. Right
to left: cast metal core, core with nylon injected
around it, nylon moulding after removal of core
(courtesy BASF).

one becomes the skin of the moulding and the other becomes the core.
The skin material can provide a high gloss or an abrasion resistant
surface, for example, while the core provides high stiffness, or low cost,
perhaps as recycled polymer. The original ICI process, later refined by
Battenfeld and by Billion, focused on applications using a foamed core
to maximize stiffness/weight ratios.

– Fusible core injection moulding. This is a route to producing hollow
 objects with a complex configuration. It is a 'lost wax' type of process, in
 which the component is formed around a low-melting metal core (e.g.,
 bismuth-tin alloys melting below 150°C). The core is inserted in a conventional tool and the plastic material injected around it (see Fig. 2.5).
 The core is then melted out, by induction heating and then drained
 away, ready to be recast into new cores. Considerable production
 engineering input has been devoted to devising economical ways of
 recovering this alloy and recasting it. The metal is expensive, and cores
 can be extremely heavy, necessitating special techniques for handling

them and locating them accurately between the two halves of the injection tool. The best-known applications are automotive inlet manifolds. After a prolonged period of development these have now been established in both thermosets (BMC) and thermoplastic (glass reinforced nylon 66).

– Injection moulding thermosets. In principle the process is the same as for moulding thermoplastics. The difference is the vital need to control the time and temperature variables very precisely during the plasticization stage, to avoid premature curing. Engineering phenolics are now well established for close-tolerance injection moulding of underbonnet components such as water pump housings. Polyester based BMC has now achieved high-visibility, high-volume use in tailgates, such as that of the Fiat Tipo.

Extrusion

Unlike injection moulding and most of the other operations, extrusion is a continuous process. Thermoplastic granules are plasticized and melted in a heated barrel and conveyed forward by means of a screw through a shaped die to produce a continuous section. The automotive products are well known:

– Cable sheathing; in the motor industry, normally in PVC.
– Monofilaments, in old established applications like nylon brush filling, and new ones like acrylic fibre optic filaments.
– Film, in basic uses like LDPE protective film in the showroom, and sophisticated applications like the biaxially oriented PET film used in instrument panel displays.
– Tubing, such as nylon 12 in fuel and pressurized air lines.
– Shaped sections, such as contoured PVC lateral protection strip.

Blow moulding

Blow moulding is the simplest and most effective method of manufacturing hollow plastic articles. There are three principal stages in the operation: melt is extruded continuously into an accumulator, tubular preforms are repeatedly ejected downwards from the accumulator by means of ram extrusion into the open blowing tool, then the tool is closed and the preform inflated. Very viscous polymer melts are necessary to ensure the form stability of the tubular 'parison' between its emergence from the die and the blowing stage. It is not a high pressure process and the tooling is relatively cheap. Inevitably, the quality of the surface tends to be inferior to injection moulding, and the inner surface is not directly shaped by the tool.

Blow moulding is not confined to hollow parts. 'Twin shell' forms can be blown as a single entity and then parted; this can still be a very cost-effective solution compared with injection moulding. Blow moulding has matured with the aid of modern computer technology. It is now possible to blow highly asymmetrical shapes without unacceptable variation in wall thickness by computerized parison control, and to locate plastic or metal inserts accurately in the tool. Multi-layer blow moulding is now a reality, offering the facility of combining the benefits of two dissimilar materials.

Expansion tanks, instrument panel sections and heater ducting are long-established blow moulding applications (in PP), but by far the most important application is the fuel tank, in high molecular weight forms of HDPE. Since its first volume application in the VW Passat of 1973, the concept has spread very widely, into some very strange custom-built shapes, literally designed to fit the space available.

Vacuum and pressure forming

Vacuum and pressure forming techniques are used to shape thermoplastic sheet around a former. Pressure forming is a more recent development, and by definition can utilize a greater differential pressure; hence it can reproduce more precise detail. They are relatively low-cost processes, both in terms of tooling and operation, and consequently are useful for limited run applications, and also for prototypes and pilot runs for high-investment injection moulding projects. Their principal use in the motor industry is for textured cladding surfaces in the passenger compartment.

Forming of SMC

The original 'glass reinforced plastics' (GRP) of the 1950s were liquid resins, epoxy or unsaturated polyester, made to impregnate a system of glass fibres (usually a random mat of E-glass fibres) before being cured. The basic process of hand lay-up is extremely labour-intensive and only gives precise forming of the outer surface; its great advantage is very low cost tooling, thus making it ideal for very short run production. Several methods have been devised for improving the surface quality by added pressure and for speeding up the process, but by far the most important contribution to the automotive industry has been the development of SMC.

SMC is very well established in the industry. With annual production varying from 20 000 to 50 000, the Chevrolet Corvette has had its body constructed from SMC panels since the mid-1960s, replacing the more primitive hand lay-up variants employed in the original 1953 model. Figure 2.6 illustrates the 1989 version, with SMC panels.

2.6 1989 Chevrolet Corvette, with SMC panels
(courtesy GE Plastics).

SMC is supplied as sheets of glass fibre impregnated with polyester resin, incorporating a catalyst and usually an added inert filler. It is moulded between heated matched metal moulds, on a press (a reassuringly familiar technique for the automotive engineer!). Production rates, though not comparable with steel pressing or injection moulding at its best, are acceptably high.

Although most SMC formulations are based on traditional polyester and random fibre mat, there are many variants available today; some are based on vinyl ester, and others include a small proportion of thermoplastics. Formulations designed for high rigidity applications like bumper beams may contain around 60% of glass fibres, most of them uni-directional.

The first significant European SMC applications were the bumpers and side panels of the Renault 5. The status of SMC as a body panel material was greatly enhanced by its success in truck cabs in the early 1970s. The ERF cab, the world's first all-plastic truck cab, was composed of 17 separate SMC panels bolted to a steel frame (see Fig. 2.7). The benefits were a two-thirds reduction in tooling costs and improved design flexibility. This has been a widely exploited method since. Later developments have included the bonnet of the Citroen BX (since 1982), and the front end moulding

2.7   ERF cab made of SMC panels (courtesy DSM Resins).

2.8   Air intakes for Ferrari 348, in SMC made by DSM Italia (courtesy DSM Resins).

(supporting the headlamps) of the 1990 VW Passat. These (the Passat especially) are high volume parts, painted on-line to a Class A finish. Modern formulations and techniques enable high-definition, heavily contoured surfaces to be produced; the air intakes of the Ferrari shown in Fig. 2.8 are an excellent example.

Forming of glass mat thermoplastics

Glass mat thermoplastics (GMT) are materials with rather different characteristics, composed of a thermoplastic matrix reinforced with a glass mat. There are two distinct production processes for GMT: one uses melt extrusion to impregnate a preformed glass mat, and the other is based on paper making, commencing with a slurry of powdered polymer and chopped glass fibres, then pressing, drying and consolidating.

Several engineering plastics have been used as the matrix in both of these GMT systems; however the versatile and modestly priced polypropylene is by far the most popular. Both types of GMT are fabricated by a process known as 'hot flow stamping'. This is similar to sheet moulding, but uses preheated blanks pressed in cold moulds, with a cycle time of less than one minute, comparable with injection moulding. (SMC uses heated press tools to mould cold blanks, with rather longer cycle times.)

Apart from faster cycling, the advantages of GMT compared with SMC are the indefinite shelf life of the blanks and better impact strength in the product. The most serious limitation of GMT is the inability to achieve Class A surface finish, which precludes the use of PP glass mat in body panels. PET and PBT based glass mat systems have been proposed for exterior body panels, with an additional layer of glass-free film to enable Class A

2.9   Radiator support panel in GMT-PP, from
1989 Chevrolet Corvette (courtesy GE Plastics).

finish to be achieved over the complete range of topcoat temperatures. Specific commercial applications had yet to appear at the time of writing.

Since the early 1980s numerous applications have been developed, such as engine under-panels, battery trays, radiator support panels and seat backs. These are all unpainted, 'unseen' parts, with PP as the polymer matrix. Figure 2.9 shows the radiator support panel in GMT-PP for the 1989 Corvette pictured in Fig. 2.6. GMT fabrication also lends itself to laminating with films or textile coverings.

High performance structural composites and prepregs

The aerospace industry is now using structural composites in significant quantities. Most of these feature reinforcing mats containing both random and uni-directional continuous fibres, with carbon fibre as the dominant reinforcing agent. Polymer matrices are usually thermosets with high temperature resistance and low flammability, sometimes toughened by additions of aromatic thermoplastics. These same materials reinforced with uni-directional carbon fibres are used to make high strength thermoplastic prepregs.

All this is a far cry from the motor industry as we know it. The aerospace industry has to meet very exacting specifications for strength and flammability, and above all, it is prepared to pay handsomely for weight reduction. The motor industry does not have this motivation, and hence these high cost, sophisticated composites are unlikely to be used beyond Formula One racing, where structures in high strength carbon fibre composites have already proved their worth. (The 1988 Williams car, for example, had a body shell of double-skinned carbon fibre reinforced epoxy, with a non-combustible core of DuPont meta-aramid fibre.)

Nevertheless, aerospace developments are relevant to the motor industry. Work is in progress in many countries, applying aerospace concepts using cheaper materials. Composite structures, based mainly on epoxy and continuous glass fibre, have been developed over many years. The best known applications are leaf springs, formed by compression moulding impregnated continuous fibre, and filament-wound drive shafts. The technical performance of these critical and complex components has been repeatedly proven, but the costs are still unacceptable for volume production. It is only by greatly increased industrialization and automation of the fabrication processes that these excellent materials will come into general automotive use.

Reaction injection moulding

Reaction injection moulding (RIM) involves bringing together two liquid monomers in a mould and causing them to polymerize into shape. The

2.10   BMW dashboard: ABS/PVC skin backed
with semi-rigid PUR foam (courtesy ICI
Polyurethanes).

process has developed around polyurethanes, in their various guises. RIM
systems with other liquid monomers have emerged, but none has approached
the success of the PUR family. RIM products differ from other plastics in
that they are not supplied as a commodity in sacks or bulk containers, and
the finished articles are created direct from the monomers.

Some 90% of the total world PUR market is in foam. Foam results from
the formation of gas bubbles in the polymerization process: originally these
were all CFCs, but the producers are moving rapidly towards chlorine-free
blowing agents. There are three distinct types of PUR foam:

1. Low-density flexible foams. These are open cell foams, excellent as
   resilient cushioning material, and universally used in car seating. Semi-
   rigid variants have good shock-absorbing properties and are much used
   in automotive trim as well as filling for bumpers. Semi-rigid PUR is the
   shock-absorbing backing material for the ABS/PVC-skinned fascia in
   the BMW fascia shown in Fig. 2.10.
2. Low-density rigid foams. These are highly cross-linked polymers with a
   closed cell structure, which are not at all flexible. Excellent insulating
   materials, they are extensively used in refrigeration, but very little in the
   automotive industry.
3. High-density flexible foams. These have a foamed cellular core, but with
   a relatively solid skin. They are increasingly important in the automotive
   industry for integrally-skinned trim.

Solid PUR elastomers are useful engineering materials, as seals, bushes,
gaiters, etc., with extraordinary abrasion resistance. Exterior car body parts
such as bumper skins and wings are actually microcellular elastomers, with
very high density foam.

The pressures generated during reaction are never more than around 5%

of those experienced in injection moulding, and hence the costs of the general equipment and the individual tooling for RIM are very much lower. Furthermore, prototype production is much cheaper and faster. Conventional injection moulding enjoys economies of scale compared with RIM, because of its higher production rates, particularly with smaller components. The difference in viability threshold between RIM and conventional injection moulding is diminishing, however, as RIM technology improves, and more reactive PUR variants such as polyurea evolve.

Flexible PUR is well suited to tough, deformable automotive parts like bumper covers and wings. Reinforced RIM (R-RIM) was developed to embrace more rigid applications, without sacrificing the valuable resilience of PUR. This entails incorporating very short (hammer milled) glass fibres into one of the monomers, at up to 15% concentration in the final polymer. The change in property profile is sufficient to secure numerous applications in small volume speciality cars; examples include the bumpers of the Lotus Excel and Renault Alpine V6, and the front and rear ends and wings of the Reliant Scimitar.

A recent development which is more spectacular in performance capability and potential application is structural reaction injection moulding (S-RIM). This involves placing a preform, usually of glass mat, in the mould, and then injecting the monomeric liquids under pressure. It is potentially a high speed process: the limitations (now being widely addressed) are in the techniques for forming and placement of the reinforcing fibres and preforms. Early applications were in rigid panels for small and medium volume models, such as the rear quarter panel for the Ford Escort Cabriolet. S-RIM is now beginning to appear in more critical, higher volume parts like seat shells and instrument panel armatures. Higher strength (compared with SMC and GMT) is claimed in uni-directional reinforced structures, because the fibres can be more precisely located.

Resin transfer moulding

Resin transfer moulding (RTM) originated in France in the early 1960s, particularly with the Matra organization, making exterior panels for small series speciality vehicles. Like S-RIM, the process consists of locating fibre reinforcement in a mould, and then polymerizing around it to yield the finished, shaped component. Rather than reacting two different liquids, RTM utilizes a partially reacted but still liquid thermosetting polymer. RTM can be used for a wide range of chemical types, and consequently a wider range of end effects than S-RIM, but inevitably the viscosity of the matrix when injected is higher, penetration is slower, and the cycle time is longer. The demarcation between RTM and S-RIM is becoming blurred, as new

Table 2.5. Processes compared

Process	Initial costs	Production rate
Injection moulding	H	1
Structural foam moulding	H	4
Blow moulding	H	3
BMC moulding	H	2
SMC moulding	M	3
GMT stamping	M	2
Rotational moulding	M	4
RIM	L	3
RTM	L	4
Thermoforming	L	3

H = High; M = Medium; L = Low.
1 = Very fast; 2 = Fast; 3 = Medium; 4 = Slow.

work is done on the chemistry of the resins and the technology of the reinforcement, producing increasingly sophisticated preforms.

Polyester resin is the most extensively used matrix; one of the most successful examples of its use is in the side panels of the Renault Espace, produced by Matra at a volume of over 200 per day. RTM has recently been used to replace aluminium in body panels for coaches. A pressurized system is used, with the reinforced polyester backed up with fire-retardant PU foam, and fabric. Several new resins have been developed, offering better performance and faster curing.

The 'numbers game' in plastics processing

Awareness of the 'numbers game' is crucially important to a designer of plastics components. The most important aspects are: (1) the total numbers required, (2) the up-front costs which have to be spread over the total production volume, and (3) the rate of production which the chosen process can achieve. Table 2.5 gives a very simplified basic comparison of initial costs and production rates for the most relevant processes.

Designing with plastics

The principles of good design are universal and applicable in some measure to all classes of material. There are however certain aspects of the behaviour of plastics which differentiate them – from metals in particular – and these no designer can afford to ignore. This section, therefore, deals with those design characteristics of plastics which are significantly different from metals.

Design freedom

Design freedom (imprecise though the concept may be) nevertheless en-capsulates most of the benefits of using plastics. These benefits can be summarized thus:

– Freedom of shape and form. Injection moulding allows all the design freedom of die casting, but with fewer finishing operations. Blow mould-ing can provide containers of very intricate shape in a single operation.
– Component consolidation. The potential for drastic reduction in the number of parts and sub-assembly operations is very important to the automotive assembly line.
– Rational design capability. Processes involving two moulded surfaces allow strength and stiffness to be controlled merely by means of wall thickness and ribbing. In the same way, fracture points can be deliber-ately designed in, e.g., for single-use trim fasteners.
– Localization of strength and stiffness. This capability is enormously en-hanced by new composites techniques for locating reinforcing mats and uni-directional fibres.
– Inclusion of inserts. 'Best of both worlds' solutions are possible by insert moulding of, e.g., brass screw threads, stainless steel bearings, steel cores for close-tolerance gears, etc. With robots the insertion can be very rapid in blow moulding, SMC and GMT processing etc., as well as in injection moulding.
– Stress relaxation. Plastics are very forgiving, in respect of load spreading. In gear teeth, for example, point contact becomes line contact, and line contact becomes area contact. Plastics under local overload tend to redistribute the stresses, allowing high (but recoverable) strain around the load point. Depending on constraints, materials can deform to accept higher loads in the stressed direction. (Reinforcement reduces this facility: very little plastic deformation is possible and the elongation to break may be only 2% to 4%.)
– Induced orientation. One unique feature of semi-crystalline plastics is the possibility of molecular orientation being induced by controlled stretching. This capability is central to the entire synthetic fibres industry built up on nylon, PET and PP over many years. Applying the principle to injection mouldings, by designing in thin sections which can be care-fully stretched, resulted in the production of the polypropylene acce-lerator pedal with a 'live hinge' in the 1960s, and later a variety of control and support devices in nylon and PP, together with 'clamshell' designs of electrical connectors.

Dimensional problems

All the dimensional problems associated with metals are present in plastics, but to a greater extent. In addition, there are some entirely new ones. All these problems are predictable and containable, but they can never be ignored. It is probably true to say that failure to allow for dimensional changes is responsible for more in-service failures with plastics than any other factor. The six prime causes follow:

1. Shrinkage. Thermoplastics generally exhibit more shrinkage than thermosets, and crystalline plastics shrink more than amorphous ones. The problem is most in evidence in injection moulding, particularly in predicting the precise shrinkage of crystalline plastics, and especially when fibre reinforcement produces orientation. Distortion results from uneven shrinkage, so inevitably the high shrinkage materials are more susceptible. As shrinkage is affected by the temperature, pressure and time settings during the moulding process, the design of moulds and the prediction of shrinkage becomes a highly skilled operation.

2. After-shrinkage. This phenomenon arises from the release of moulded-in strain. What happens is that further heating after removal from the mould is found to cause additional shrinkage, particularly in thick section parts, and distortion will appear in parts with varying wall thickness. Post-shrinkage is a 'one-off' effect, which can be corrected by annealing, i.e., controlled re-heating. Annealing should be carried out at a temperature some 10–20°C higher than the highest temperature expected in use, and rapid heating or cooling should be avoided.

3. Anisotropy. The addition of fibres to a matrix, especially if processing involves high speed melt flow, as in injection moulding, inevitably results in anisotropy. This is the directional dependence of properties and dimensions; a fact of life with fibrous composites which cannot be ignored. Orientation in the melt causes differential shrinkage, visible as distortion, and this is likely to be exacerbated by an increase in temperature. The effects can be minimized by correct tool design. Crystalline plastics are more susceptible than amorphous, because the general level of shrinkage is higher, and thermosets are the least affected of all. Anisotropy is less of a problem in SMC and GMT, and in S-RIM and RTM the dimensions are dominated by the glass mat insert.

4. Thermal expansion. Coefficients of thermal expansion are frequently represented by a single figure. This is inadequate for plastics in general, and for composites it is positively misleading. Because molecular movement increases with temperature, so does the expansion coefficient. In fibrous composites, the coefficient at any temperature is directionally dependent: unfortunately in tabulated data the figure presented is nor-

mally that measured in the most favourable direction on standard test-pieces. The actual expansion shown by a composite in any one direction is a function of the degree of orientation and of the section thickness. In practice it is also influenced by physical constraint; a component will expand more along the directions where it is free to expand. Once again, crystalline thermoplastics respond much more to temperature changes than amorphous, and some thermosets are barely affected at all.

5.   Absorption. Absorption of fluids by plastics is accompanied by a dimensional change. This can be an indication of incipient deterioration, as with many amorphous plastics immersed in organic solvents, or with some filled and laminated thermosets immersed in water. Rather different is the case of nylon, in which the polymer absorbs water only until it achieves equilibrium with the ambient humidity. This process is precise and predictable, but allowance must be made for it.

6.   Creep. Dimensional changes due to creep must always be expected.

Enhancing rigidity

The conventional methods of improving the rigidity of a component – ribbing, box sections, etc. – are all applicable to plastics. These methods can be less satisfactory than with metals: rates of cooling and total shrinkages in moulding are affected by wall thickness to a marked extent, especially in semi-crystalline plastics, so that the appearance of the surface can be impaired. The modified processes described earlier (structural foam, sandwich moulding and gas injection) all offer the prospect of increasing component stiffness, along with their other benefits. The gas injection processes, apart from the possibility of providing internal channels, can give components having both flexible and rigid areas, without the distortion and surface blemishes usually associated with varying wall thickness.

The precise location of reinforcement, as continuous fibres or mat, is increasingly seen as a way of controlling rigidity in all the moulding and forming processes. Blow moulding has also been used to provide stiffness without resort to glass reinforcement, in bumper beams and indeed in bumpers themselves.

Creep deformation

Creep is an extremely important concept in plastics. Deformation is in fact a much more common cause of failure in plastics than it is in metals. Awareness of creep is important with plastics at any temperature, whenever the material is under load. Design calculations involving a modulus should use a

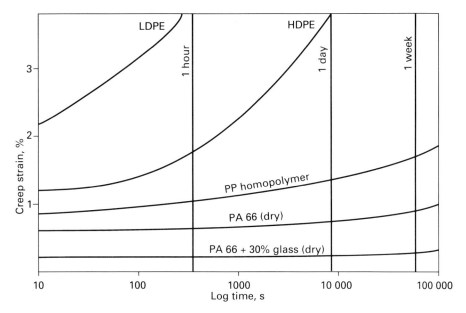

2.11 Creep in various polymers compared at distant (20°C) temperature and load (20 MPa) (data supplied by ICI).

value which is appropriate to the temperature and time under load. Hence, we have the concept of the tensile creep modulus.

An explanation of the importance of creep modulus is appropriate. Loading produces a deformation, described as creep strain. With continued loading the movement continues, but with progressively smaller increments. (As an example of this time effect, consider the case of a nylon nut on a steel bolt. The nut, tightened to a specific torque on assembly, may need to be retightened after, say, two days. Further creep will occur much more slowly, and it could be another year before a similar loss of torque is experienced.) Creep strain grows with higher temperature and time under load. In practice, creep is more likely to be encountered in flexure or compression, but laboratory measurements are normally carried out in tension, which allows much more accurate monitoring. What is actually measured is the extension, i.e., the strain increase with time. For composites, the direction of the applied stress in relation to the fibre orientation must be specified (creep resistance being much higher in the fibre direction). Figure 2.11 shows creep curves for different polymers at the same temperature and load.

Creep measurements can be converted into more useful parameters, and

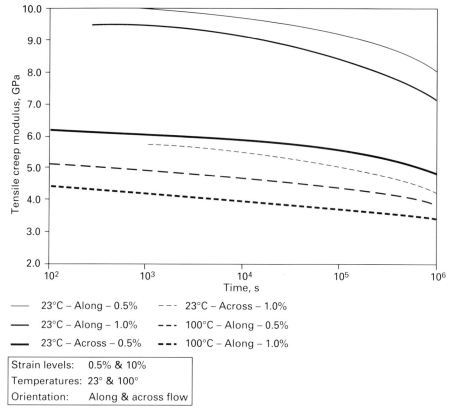

—— 23°C – Along – 0.5%	– – – 23°C – Across – 1.0%
—— 23°C – Along – 1.0%	– – – 100°C – Along – 0.5%
—— 23°C – Across – 0.5%	– – 100°C – Along – 1.0%

Strain levels:	0.5% & 10%
Temperatures:	23° & 100°
Orientation:	Along & across flow

2.12   Factors affecting tensile creep modulus: 33% glass fibre reinforced PA66 (data supplied by ICI).

in particular the tensile creep modulus, by relating the measured strain to the applied stress. Curves of creep modulus against time are directly useful in design calculations. Methods have been devised for interpolating detailed families of curves from a few accurate measurements.

Figure 2.12 shows how the apparent modulus (the tensile creep modulus) decreases with time under load. In this family of curves for a single grade of glass fibre reinforced nylon, it can be seen that the apparent modulus is reduced by temperature rise, by increased load, and by applying the load in the direction in which the fibres provide least support, i.e. at right angles to the fibre direction. Families of curves of this kind are available from the polymer suppliers and the databases; they are essential equipment for the designer.

Avoiding brittle failure

The fracture behaviour of plastics is a complex subject. In the early years, when testing standards were more rudimentary, information was inconsistent and understanding was sparse. This all helped to build up a bad reputation for plastics. Improved test methods and sophisticated theories have transformed the situation, but in the popular imagination the memory lingers on.

The aim of this section is to identify the key factors controlling the toughness and brittleness of plastics, and to indicate how catastrophic failure of components can be avoided. In design terms there is a difference of approach between toughness and stiffness. There are many ways, as we have seen, of enhancing the stiffness of a component, but with toughness (an inherent material characteristic) the overriding consideration is to avoid the many errors which can impair it. Impact strength is a measure of a material's ability to absorb received energy and convert it into internal energy (roughly considered as molecular movement). If the impact strength is high, the material is described as tough. When a material is unable to absorb the external energy, so that it cracks or shatters under impact, it is described as brittle. With polymers certain background facts need to be understood.

Firstly, different impact tests give different assessments of materials. Test specimens are designed to highlight a particular effect: thus it is that any test measures the impact resistance of the specimen, and not that of the material itself.

Secondly, under different conditions, a particular test specimen can appear as either brittle or tough. Brittleness can be induced either by low temperature or by a high strain rate. At low temperature, there is not enough capability for molecular movement to absorb the impact energy, whereas at

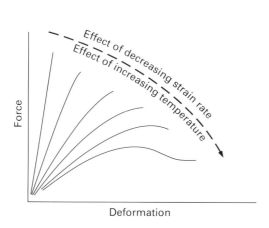

2.13   Force-deformation curves.

a high strain rate the duration of the impact is too short for molecular rearrangement to accommodate it. All materials, whether normally tough or brittle, undergo a tough-brittle transition as the temperature is reduced. Speeding up the strain rate induces a similar tough-brittle transition. These effects are illustrated schematically, by reference to the shape of the force-deformation curves, in Fig. 2.13. The actual transition from tough to brittle is rarely very precise. Usually there is a zone of variable behaviour between the clearly-defined extremes.

The third factor, which can affect this transition and make a tough material behave in a brittle way, is stress concentration, from 'stress raisers'. These are the things which cause unwarranted failure and are invariably due to mistakes by the material suppliers, or the designers, or the processors, or even the assemblers. Examples include:

- Poorly dispersed pigment.
- Dirt.
- Voids.
- Sharp angles.
- Sudden changes in thickness.
- Surface cracks, scoring, or excessive texturing.

On occasion a stress raiser is unavoidable, as with moulded-in metal inserts. The designer must then ensure that stress concentrations are minimized, and that there is an impact strength safety margin. A common design fault is the unseen internal angle. As a guideline, the toughness of a part with an unradiused internal right angle will be approximately doubled by giving that angle a 0.5 mm radius.

The effect of stress raisers is simulated in the laboratory by evaluating impact strength in specimens which have been notched to different degrees of severity, assessed by the notch radius (respectively 0.25 mm, 2 mm, and unnotched). This is demonstrated in Fig. 2.14, which compares a 'tough' material, PA 66, with its reinforced version containing 33% glass, which can be described as 'strong, brittle'. Figure 2.14 demonstrates (a) the beneficial effect of temperature on impact strength, (b) the deleterious effect of notching on impact strength, and (c) that the reinforced polymer is much less notch sensitive (and temperature dependent) than the unreinforced.

There are different forms of brittle failure. Figure 2.15 shows specimens impacted on a falling dart test. On the left is a 'tough' unreinforced material tested under extreme conditions, having shattered catastrophically. On the right is a short fibre reinforced composite, with a much higher resistance to crack propagation, failing after repeated blows. The longer the fibres and the more effective the reinforcement, the more the component is likely to

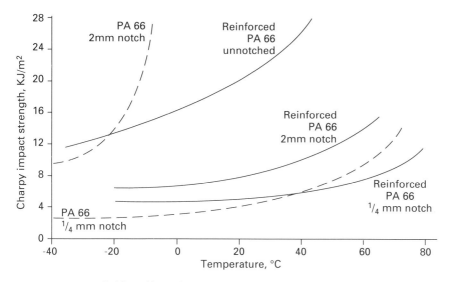

2.14   Effect of temperature and notching on impact strength (data supplied by ICI).

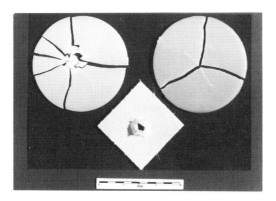

2.15   Types of brittle failure, in falling dart impact test (courtesy ICI).

retain its form after failure. The lower picture shows a sample of glass mat reinforced material, which has retained its integrity even after failure.

Composites also behave differently from unreinforced plastics in their response to an embrittled surface. A specimen of polymer which shows tough failure, i.e., by irreversible plastic deformation, can be made to fail in a brittle way simply by acquiring a brittle surface. This can result from oxidation, by heat, or ultra-violet (UV) light or radiation, or from being painted with a brittle coating (as when a bumper shell in a tough plastic is on-line painted with a hard, brittle topcoat, without an intermediate flexible primer). Again, because of its higher crack propagation resistance, a rein-forced specimen will retain its integrity even when it is cracked by an impact.

Another factor affecting toughness is the length of the polymer chains, expressed as molecular weight. Not surprisingly, it is the higher molecular weight grades of polymers which deliver the best impact strength, as in the fuel tank grades of HDPE. Factors like heat ageing and exposure to aggres-sive chemicals, by reducing the average molecular weight, will eventually induce brittleness.

Further aspects

There are many other aspects of plastics behaviour with a bearing on practical design, which can be noted only briefly here:

- Creep rupture. Metals under load will in general deform elastically, recovering immediately when the load is removed. In plastics, this situa-tion prevails only at very low load. Additional loading induces a degree of molecular rearrangement, or plastic deformation. Some of this is recoverable, but a point is reached where the movement becomes irre-versible. The visible effect in a tensile test piece under excessive creep loading is that eventually necking begins at the weakest point, and the creep strain increases rapidly until rupture occurs. The higher the load (and the higher the temperature), the sooner the onset of necking, followed by creep rupture.

- Stress redistribution. More realistic than the 'pure' tensile loading is the case of a component subjected to a mixture of compressive and flexural loads, with constraints in certain directions. Under these conditions, the part will accept higher loads than theory would predict, although this may be accompanied by high deflection.

- Varying or intermittent loads. The essential point is that the pace of molecular rearrangement is relatively slow. Creep strain under load rises quickly at first, and then more slowly; when the load is removed, the material begins to recover. Recovery is fast at first, and then slows down progressively. Because of recovery, intermittent loading produces less strain then a steady load, providing always that the strain remains within the defined 'design strain'.

- Dynamic fatigue. Intermittent or cyclic loading can cause two additional problems. Because of the low thermal conductivity of plastics, high frequency cyclic loading can cause internal heating, with consequent softening of the material. However, even at low frequency cycling, repeated intermittent loading can produce brittle fracture. This is the phenomenon of dynamic fatigue, which really arises because the 'natural' process of molecular rearrangement around the stressed points is inhibited by the continually changing stress pattern.

Components in practice

Real life situations rarely match up to the 'pure' laboratory conditions: nevertheless, much can be done with 'informed approximations', provided we are aware of the principal factors at play. Certainly we should be able to dispense with the traditional blind method of 'data sheet less safety factor', and determine a design stress appropriate to a specific material in a particular application.

One very effective way of doing this is the 'maximum strain' approach. The starting point is to nominate the maximum strain which a component can tolerate, from knowledge of its intended function. The appropriate design stress is then found by referring to the isometric stress/time curves, derived from creep data. Nowadays curves are available for all the best-known materials; they are plotted as a family of stress/log time curves at various strain levels. Some experience is necessary in nominating the appropriate design strain; for homogeneous components in a tough plastic a figure between 2.5% and 3% is typical, but for welded joints and for reinforced plastics, the figure should be reduced to 0.7% to 1%. The lower figure of each pair is appropriate when the performance conditions are not completely predictable.

The stress/log time plot can be used to convey a great deal of information about a material. Figure 2.16 is a design plot which indicates on the same coordinates (1) isometric stress/log time curves at specific strain values, (2) the creep rupture curve for the material, derived by measuring the times taken for samples to rupture under various constant stresses, and (3) a dynamic fatigue curve obtained by measuring the times to failure under cyclic on/off stresses at various stress levels. Although (2) and (3) are not strictly comparable with (1) because the conditions of experiment are necessarily different, the design plot does give a good overview of a material's capability, and gives a realistic assessment of the safety margin at the selected design strain.

The foregoing gives some inkling of the pitfalls in applying Finite Element Analysis (FEA) design techniques to plastic components. Ideally, FEA should allow for the non-linear behaviour of plastics at high strain: a really rigorous analysis would allow for recoverable strains (in a tough material) of up to 3%, at which the modulus may be only 60% of its low-strain value. Some FEA programmes enable this to be done, but the processing time and

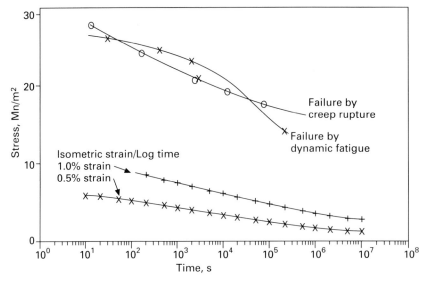

2.16   Design plot, composed of stress/log time curves, for a PP copolymer (data supplied by ICI).

costs are excessive. However, simple low-strain programmes are still of value, because they will suggest the most effective distribution of stress levels, even if the ultimate load capabilities are not given.

Components often include discontinuities which are not present in the testpieces from which performance data are derived. Welded joints are an obvious example; apart from the incipient weakness to which all materials are subject, reinforced materials are especially vulnerable, because there will be no fibres crossing the boundary. In injection moulded components, weld lines formed where two streams of melt have met are always likely to display weakness; again, reinforced materials are more susceptible.

Figure 2.17 is a representation of how design faults can cause a component to fail a specification. For example, parts in the engine compartment may be required to perform between −40 and +150°C. The limitations which really define the useful working temperature range of a material are its low temperature impact strength and its high temperature stiffness. Hence to be approved against the relevant manufacturers' engineering specifications, a material may be required to pass in turn tests which measure impact strength at −40°C and modulus at 150°C. The solid curve in Fig. 2.17(a) shows the ideal impact/temperature curve for this material, achieving the required impact strength (A) at −40°C. If, however, because of design or processing faults the effective real impact/temperature curve is the dotted

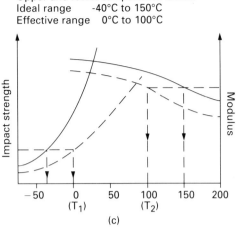

Upper and lower temperature limits:
Ideal range   -40°C to 150°C
Effective range   0°C to 100°C

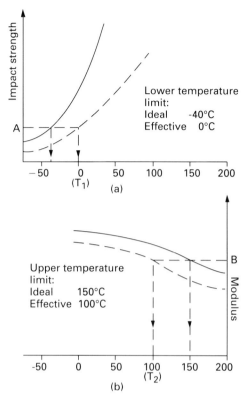

2.17   Schematic representation of effect of design faults on temperature limits: (a) low temperature impact, (b) high temperature impact, (c) the summation.

one, then the specified impact strength (A) is only achieved at a higher temperature T1. Modulus can be treated similarly: the solid curve in Fig. 2.17(b) is the ideal modulus/temperature curve, showing the specified modulus (B) being achieved at 150°C. However, if the component is operating in a creep situation, or if it is a reinforced material unfavourably orientated, then the effective curve is now the dotted one, where the specified modulus is sustained only up to a temperature T2, instead of 150°C. Putting the two curves from the two parameters together in Fig. 2.17(c), we see that the effective working temperature range of the component is now only T1 to T2, although the material is capable of functioning over the complete range from −40 to +150°C.

Coexisting with metals

Often, in the real world, the best solution to a design problem will involve plastics coexisting with metals. Awareness of the problems of such coexistence can be vital to the design of components for sub-assemblies:

- Differential expansion. Ignoring this can have serious consequences. One solution, when other needs permit, is to use unreinforced thermosets, with extremely low expansion. The early attempts in the USA to use glass reinforced nylon for rocker covers highlighted the problems: the composite not only expanded more than the metal, but its expansion increased with temperature, and furthermore varied across the moulding, being different around the moulded-in bolt holes, where the fibre orientation was different. Inevitably, oil leaks occurred between the bolt fixtures. Successful rocker covers have been designed, using lower expansion composites based on unsaturated polyester (as DMC or SMC) and (in the case of the Citroen AX) on semi-aromatic nylon. The biggest single factor however is the redesign of the gaskets to cope with differential expansion. The success of the glass reinforced nylon radiator tank also owes much to the generously-proportioned synthetic rubber gasket which marries the aluminium radiator to the composite cover, and absorbs the dimensional changes.

- Anisotropy. Apart from new gasket design and materials, the important aspect is to design out the distortion tendency. This can be done either by making the shape as symmetrical as possible, thus minimizing orientation, or (in mouldings with a high aspect ratio) deliberately inducing high but uniform orientation by gating at one end. The aim should be to avoid a situation where the orientation can vary with the processing conditions.

- Metal inserts. This is a good means of overcoming some of the deficiencies of plastics. Metal inserts are helpful in many ways: (1) to reduce the overall thermal expansion of a component, (2) to increase rigidity, (3) to facilitate repeated assembly, (4) to increase the pull-out strength of self-tapping screws, (5) to provide fixture points for sub-assemblies, and (6) in bearing applications to give higher PV values and higher temperature operation. Putting robots to use in locating inserts in the mould or the press has made the operation much faster and more reliable. Inserts are now widely used in the processing of SMC, GMT, blow and rotational moulding, as well as in injection moulding.

- Moving parts. There are many systems of gears and bearings where a combination of metal and plastics parts results in a better assembly than either could achieve alone. The main benefits of plastics are in reduced maintenance, quieter running, and (in low load systems like windscreen wipers and speedometers) longer life. Plastic gears for the automotive gearbox might be unlikely, but the benefits of using reinforced nylon and PES for the bearing cages therein are widely recognized. Some very sophisticated composite plastic bearing materials are now available, using internal and surface-migrating lubricants in tough matrices reinforced with fibres; these often out-perform the most expensive metal bearing compositions.

Joining plastics

The effectiveness of plastics and composites is often critically dependent on joining technology. In the automotive industry, the key question is not only whether the joint is consistently sound, but whether making it is compatible with a high speed assembly operation. Specialist advice should always be sought in this area, because technology is changing fast, and comparisons may need to be made between quite different systems. The prime source of information is The Welding Institute (Abington, Cambridge, UK), who have expertise on adhesives as well as welding techniques.

The performance of today's highly sophisticated adhesives can be very specific, not only in terms of the materials to be joined, but also with reference to the design of a joint and its intended requirements. There are two very useful databases to facilitate the selection of adhesives: 'EASel', developed by the Design Council, and 'PAL' from Permabond Adhesives Ltd (Eastleigh, Hampshire), which is not confined to adhesives supplied by Permabond.

Designers need to be aware of problems and opportunities in joining plastics. The points peculiar to plastics are summarized here:

- The potential problems of using plastics have been highlighted in this chapter. The most common ones result from creep and relaxation, and the differential thermal expansion between different materials.
- The many advantages of plastics include their greater ability to adjust to loadings and dimensional variations. Lower softening points mean that thermal welding is easier, providing oxidation can be avoided.
- Modern quick-acting adhesives are finding their way on to the assembly line, with robots, in traditionally difficult operations like windscreen insertion. Surface preparation is vital with plastics as with other materials.
- Different rates of melting and different viscosities in the melt mean that individual plastics respond differently to the common welding methods. All of the following systems are in use:

    - Hotplate welding.
    - Friction welding (spin and rotary).
    - Ultrasonic welding.
    - Vibration welding.
    - Induction welding.
    - Hot gas welding, etc.

- Mechanical assembly methods are widely used, but there is always a risk of high local stresses. Moulded-in inserts are generally better than self-tapping screws in the more demanding applications. Sprags and staples are used very successfully with tough plastics.
- Much can be achieved with push-fits, especially when a resilient material is being attached to a more rigid one. Design features can be included which are similar to those employed very successfully in nylon clips and fasteners.
- Having focused for a long time on how to make plastic assemblies secure and permanent, designers are now having to ensure that they are easily dismantled, for segregation, recovery and recycling. This turnabout is necessitating a complete rethink of all assembly operations. In some cases, the two objectives will be incompatible.

3

Choosing plastics

This chapter explores the influences leading to the use of particular plastics in particular applications. The first section identifies the characteristics which are decisive in motor industry applications, the second summarizes the strengths and weaknesses of the different polymer groups, the third offers suggestions for a materials selection procedure, and the fourth looks at the requirements of the different application areas. This is intended to provide a logical background to the detailed application surveys in Chapters 4, 5, 6 and 7.

The decisive properties

Introduction

The process of selection between different plastics can be quite bewildering. Copious information on materials is available from the manufacturers, but traditionally this tends to play down any reference to the defects, and to omit any meaningful comparison with alternative plastics. Nowadays there is a proliferation of databases, offering a facility for computerized selection from an impressive list of materials. Data have been compiled over a very wide range of properties, often with an unbiased presentation and uniform test methods and units. With such a huge bank of data now available, the problem is frequently reduced to being one of deciding which of the abundant properties really matter for the component in hand.

Temperature resistance (short term)

The common indicator of the short term upper temperature limit for a plastic material is the heat distortion temperature (HDT). Measured at two

3.1   Typical modulus temperature/curves for crystalline and amorphous plastics, with and without reinforcement (data supplied by ICI).

arbitrary loadings, it is a good practical guide to the suitability of materials for components working under load, although it has no fundamental significance. A more fundamental characteristic is the glass transition temperature (Tg), which is the temperature at which a significant loosening of the internal structure occurs and molecular movement becomes possible. Above the Tg, amorphous plastics soften and no longer have any load bearing ability; crystalline plastics experience a drastic reduction in modulus, but retain their form stability, up to the crystalline melting point (Tm). Fibre reinforcement can enable crystalline plastics to be used in load-bearing applications far above their Tg, close to their Tm. Glass reinforced nylon and polypropylene components in the engine bay are frequently operating in this zone between Tg and Tm. Figure 3.1 compares modulus/temperature curves for two chemically similar amorphous and semi-crystalline plastics, with and without reinforcement.

At some point above their softening and melting points, thermoplastics begin to degrade and will eventually decompose. Thermosets are different in two respects. Exposure to high temperature can increase the cure, leading to further cross-linking, which has the effect of raising both the HDT and the Tg. Decomposition does occur at higher temperatures, taking the form of charring without melting. With some thermosets even exposure to temperatures around 1000°C still leaves a bulky carbonized residue (which explains the use of thermosets as heat shields in space travel).

It is not only the mechanical properties of plastics which are sensitive to temperature. Most characteristics are adversely affected by temperature; particularly relevant to the motor industry are chemical resistance, electrical properties and dimensional stability.

Long term temperature stability

The problem as regards long term temperature stability is thermal ageing. Plastics when exposed to heat in air (or other oxidizing media) suffer progressive deterioration through embrittlement of the surface. The effect is cumulative, and (unlike the short term temperature effects) irreversible. Because oxidation proceeds inwards from the surface, thin sections are more susceptible than thick ones. Just how serious the effects of oxidation are depends to a great extent on the crack propagation resistance, which is why fibre reinforced plastics tend to survive longer than unfilled materials, especially if the reinforcement is present as long fibres or glass mat. For components located close to heat sources, such as bulb holders and most under-bonnet parts, many formulations will include a so-called 'heat stabilizer', which delays the oxidation process. These additives are particularly beneficial in unfilled plastics, which tend to have poor resistance to the propagation of cracks initiated in an embrittled surface.

There is an approximately straight line relationship between temperature and the log of time to degrade; raising the temperature by 10°C roughly doubles the rate of oxidation. Several indicators of durability at high temperature are in use. It is customary to speak of 'continuous service temperature' without bothering to define it. This is a bad practice, and unnecessary, because the Underwriters' Laboratory Index system does allow for precise definitions. Indices are quoted for specific properties at three thicknesses (0.8, 1.6, and 3.2 mm), and an Index is defined as the temperature at which just 50% of the nominated property survives the stipulated exposure time. For components in passenger cars, whose active life at high temperature is typically only 5000 hours, the most relevant UL Index is the 50% temperature for 3.2 mm tensile specimens exposed for 5000 hours.

Figure 3.2 brings together the two concepts of short term and long term temperature limits in a single graph, for a number of automotive engineering and specialist polymers. Heat distortion temperature is plotted against con-

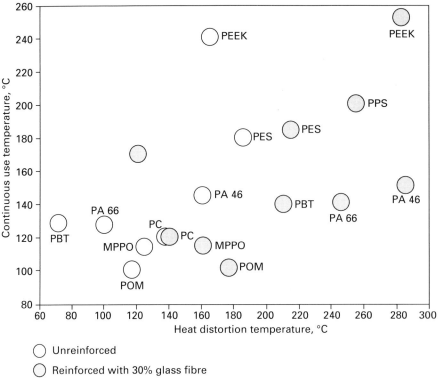

3.2   Short term and long term temperature
limits: selection of automotive plastics.

tinuous service temperature. Figure 3.3 is a generalized summary comparing the short and long term temperature characteristics of the main categories of plastics. In general, reinforced materials of all kinds tend to register the highest heat distortion temperatures, but to achieve a high continuous use temperature it is necessary to look to the aromatic and heterocyclic materials – which are also the most expensive.

In practical terms this means that the typical engine bay requirement of survival at around 130°C for 5000 hours is too severe for an unstabilized, unreinforced version of an engineering plastic like nylon, but acceptable for glass fibre reinforced nylon (e.g., in hot air intakes), and for heat stabilized unfilled nylon (e.g., in oil filler caps, cable ties and electrical connectors). Most reinforced nylon types will withstand much higher temperatures for a limited period, but for temperatures above about 130°C to be sustained for the whole life of the engine, it is necessary to go to more expensive materials, e.g., special polyamides like nylon 46, or aromatic thermoplastics like PPS or PES, or to high performance thermosets.

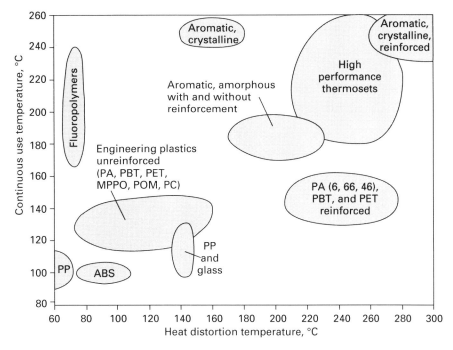

3.3   Short term and long term temperature limits: summary.

Flammability

Polymers, composed of long molecules based on carbon-carbon links, are inherently combustible. Many, such as the polyolefins, have a much higher calorific value than coal. It is not surprising that the materials suppliers have invested enormous resources into improving the flammability resistance of plastics. There are many ways of assessing burning performance, e.g.:

- Ease of ignition.
- Rate of flame spread.
- Duration of burning.
- Oxygen content needed to sustain burning.
- Presence of burning drips.
- Retention of form stability.
- Evolution of toxic gases.
- Smoke evolution.

Individual user industries attach different priorities to these criteria. The aircraft industry clearly attaches great importance to smoke and toxicity arisings. The automotive industry, however, is much less sophisticated on the subject of flammability. Specifications rest mainly on the simple concept of the rate of flame spread being slow enough to allow the passengers to leave the vehicle.

This is well founded. Evolution of smoke and toxic gases are not the most serious problems in the burning car situation; direct physiological damage due to breathing hot air and contact with flames are more immediate hazards, and (as in most fires) the most likely hazard is asphyxiation by carbon monoxide because of burning taking place under conditions of oxygen deprivation.

Although for these reasons smoke and toxicity are not particularly relevant to the choice of materials for car interiors, the designer should be aware that polymers, whether as shaped plastics or fabrics, do vary with respect to these considerations. Halogens, whether for example as chlorine in PVC or bromine in flame retardants, will produce toxic fumes above certain temperatures. Nitrogen-containing polymers such as nylon and polyurethane will (like wool) yield cyanide traces at very high temperature. Smoke can arise from the very additives included to reduce ignition and flame spread in both thermoplastics and thermosets. The best performers, in terms of low smoke evolution, are the fully aromatic high temperature thermoplastics.

Component design is also relevant. Wall thickness and lamination are important influences on the ease of ignition, and high surface to volume ratios and the presence of foam will increase the rate of flame spread.

Thermosets have an inherent advantage over thermoplastics, because they do not melt. This is the basis of the preference some manufacturers have for phenolics in the engine compartment; under extreme conditions resulting from coolant dump, for example, vital components will retain their integrity, instead of melting with even more serious consequences. However some thermosets do release inflammable volatiles at high temperature, and there may be excessive smoke evolution. Unsaturated polyester fire retardant grades perform well against burning specifications, but they contain halogens, which can lead to smoke and toxic gases. Vinyl ester variants are better in this respect.

The general guideline for thermoplastics is that the greater the aromatic content, the greater the inherent fire resistance. Hence the 'fully aromatic' thermoplastics (PPS, PES, PEEK, etc.) perform well on all the tests, without needing extraneous additives. Next in line are the 'semi-aromatic' engineering plastics like PC, PBT, PET and the semi-aromatic nylons. Other nylons and polyolefins are less satisfactory. Nevertheless, nylon 6, 66 and 46 can be formulated into halogen-free and phosphorus-free grades which meet all the electrical appliance specifications.

Chemical resistance

Chemical resistance is not easily measured in any precise quantitative way. The only information available to the designer is of a very broad qualitative nature, but a few generalizations can be made about the chemical resistance of plastics:

- Chemical resistance is temperature sensitive. If a plastic material shows any absorption or any visible change at room temperature, then there is likely to be a much more drastic effect at, say, engine compartment temperatures.
- PTFE is the most completely inert plastic, followed by the fluoro-polymers which have been modified to render them mouldable.
- Plastics based on aromatic or heterocyclic systems have very stable structures, which are very resistant to chemical attack, just as they are resistant to high temperature. The 'pure' thermoset polyimide has the greatest resistance, but the act of modifying it to make it processable inevitably impairs the chemical resistance somewhat.
- In any group of thermoplastics of comparable price range, the semi-crystalline are more chemically resistant than the amorphous.
- One of the factors making chemical resistance difficult to measure is the effect of stress. Amorphous plastics are particularly susceptible: specimens which are apparently resistant to a certain chemical will crack when a stress is applied.

For the automotive designer, there are many chemical 'hazards':

- Petrol, with and without aromatics or alcohols.
- Diesel fuel.
- Oil, and its residues.
- Hydraulic fluids.
- Hot antifreeze solutions (of different specifications).
- Battery acid.
- Cleaning solvents.
- Paints and paint primers.
- Adhesives.
- Road salt (including calcium chloride).
- Exhaust gases.

In all cases there are different degrees of severity, depending on whether the contact is total or intermittent, and whether the component is under stress.

The related topic of permeability is also very relevant for fuel tanks and fluid reservoirs.

Friction and wear

The behaviour of plastic surfaces in contact with other surfaces in motion is extremely important. Unfortunately, none of the key parameters like surface hardness, coefficient of friction, and abrasion resistance is an unchanging material characteristic. The common test methods are all arbitrary, and furthermore the surface hardness upon which the others depend varies according to the method of preparation, especially with semi-crystalline plastics. Data comparing the abrasion resistance or frictional coefficients of different material combinations under identical conditions can of course be very useful. However, data of this kind when quoted in isolation can be misleading.

Nevertheless, the first real progress in metal replacement by plastics was made in this area, when it was realized that some plastics exhibited very low coefficients of friction and possessed a natural lubricity, enabling them to operate as moving parts without external lubricant. Acetals and the nylon family are outstanding in this respect, and are well established in such applications as steering gear joints and rack-and-pinion bushes. PTFE has the best low friction performance of all, but having inadequate strength and creep resistance as a pure solid, it performs very much better in practice when dispersed within a structure of metal or another plastic. As explained in Chapter 2, the most effective bearing surfaces are proprietary composites, made up of matrix, reinforcement and lubricants.

A different problem pertains with interior and exterior automobile surfaces, where the need is to preserve the appearance after minor impacts and abrasions. Such surfaces are usually textured, and careful formulation is needed, particularly in relatively soft materials like polypropylene, to preserve a uniformly matt finish and a uniform colour.

Impact strength

Undoubtedly, impact strength must feature in the selection process. The problem is that the commonly used test methods have little relevance to real situations. The Izod and Charpy methods as normally used call for specimens which are severely notched (the sort of notches which would disgrace any designer who inadvertently included them in a component!).

The falling weight impact strength methods are much more relevant to car panels, because they measure the energy absorbing ability of large un-

3.4 Crash testing of GE Plastics 'Vectra' concept car, at Motor Industry Research Association facility (courtesy GE Plastics).

notched areas. However none of these methods has any absolute validity, and they are most effective when used in a comparative way alongside well-known standard materials. The section in Chapter 2 on brittle failure emphasized that impact performance is sensitive to temperature and the rate of application of load, and above all to the presence of stress raisers. It is perfectly possible for panels in different materials to be ranked differently at different temperatures, at different strain rates, and with different fixing arrangements.

Impact strength data can help with the pre-selection process, but beyond this should be used with great caution. Accurate prediction of collision behaviour is extremely difficult, and likely to remain so. Crash testing of vehicles shows that the actual results of collisions depend on a complex mix of design and construction factors, as well as material characteristics. It is an expensive operation, as Fig. 3.4 makes clear.

Resistance to ultraviolet light

Resistance to ultraviolet (UV) light is a matter of concern in the choice of materials for bumpers and lamp clusters, and 'hang-ons' like door mirrors. It emerges from time to time with stylists' innovative attempts to escape from the cost penalties of the paint shop, like vinyl roof covers.

Often the most obvious effect of exposure to sunlight is colour fading: this is essentially related to the pigments or dyestuffs, and not to the polymer. The other effects are the results of surface oxidation of the polymer, in a similar way to heat ageing. The first visible effect is discoloration, in the direction of yellow and then brown. Micro-crazing develops concurrently, but may not become evident until a general loss of gloss is apparent. How quickly this surface embrittlement leads to mechanical failure will depend on the strains to which the surface is subjected.

If the weather profile involves rain as well as sunlight, then the micro-crazing will be followed by rain erosion. 'Chalking' will become obvious on the surface of polymers formulated with white pigments, and in fibre-reinforced grades the underlying fibrous structure will become visible. Although the reinforced materials lose their good looks sooner, mechanical deterioration takes much longer, because of their resistance to crack propagation. The process of UV ageing can be delayed or slowed down by added stabilizers, and (where appropriate) by carbon black. Without these stabilizers, the appearance of most plastics deteriorates significantly within a few months (and much faster than automotive paint finishes). Apart from PVC (not much seen in car exteriors nowadays), the automotive polymer with the best weathering performance is polymethyl methacrylate, almost universally used for rear lamp lenses. Even then, careful pigment formulation is necessary for colour stability; attempts in the 1960s to cut costs by lacquering clear acrylic mouldings instead of compounding the polymer with pigment proved to be disastrous, because of colour fading.

Dimensions

The various influences affecting dimensional behaviour have been described in Chapter 2. It is evident that in service this vitally important parameter cannot be predicted with total precision. There are simply too many factors which depend upon design and shape. The compensating factor for all this uncertainty is that, in general, plastics components can function with tolerances which would be unthinkable with metals. The simplistic summary is that thermosets are the most dimensionally stable category, followed by amorphous thermoplastics, with crystalline thermoplastics as the group most susceptible to dimensional change. In all categories, fibrous reinforcement reduces the general level of variation, and can produce a striking improvement. Due regard must always be paid however to both the benefits and the penalties of anisotropy.

The problems likely to be met with particular materials can be confidently forecast, but the final decision on dimensional acceptability will often demand measurements on authentic prototypes.

Transparency

Occasionally in component design, transparency can be a very influential property. Total, glass-like clarity can only be found in amorphous thermoplastics. The most widely used of these are PC, PMMA, SAN (styrene acrylonitrile) and PS, of which only the first two, polycarbonate and acrylic, are really important in the motor industry. As mentioned previously, PMMA is used almost universally for rear light lenses, by virtue of its optical and ageing performance and UV and scratch resistance. Superior impact and heat distortion performance enable PC to qualify for headlight lenses and under-bonnet covers, in spite of its higher price.

Translucency, as distinct from transparency, can be achieved in a great variety of plastics if the specimen is thin enough; even in some reinforced thermosets. Examples of the application of this characteristic are PP fluid reservoirs, in which the level can be read, and interior lamp lenses in nylon, providing diffuse illumination.

Reducing the section thickness to the level of foils and films allows many more polymers to become transparent. Although films technology is very highly developed nowadays, it has only a few points of contact with automotive design, e.g., in the use of oriented polyester film for membrane touch switches and instrument panel display.

Characteristics of the polymer groups

The enormous variety of plastics, and the complex property balances attainable by formulation, make it impossible to present a summary which is at the same time compact and meaningful. The general performance indicators

Table 3.1. Cost and temperature indicators for the main polymer groupings

Group	Principal members	Cost (relative to PP = 1)	Continuous service temp, °C
1 Bulk polymers	PP, PE, PS, PVC	1–1.5	60–100
2 Bulk thermosets	PF, UF, UP, PUR	1–2	100–160
3 Transitional	ABS, SAN, PMMA	1.5–3	90–110
4 Engineering plastics	Standard PA, PC, POM, reinforced PP, mPPO	2–4	90–130
5 Engineering thermosets	EP, PUR, PF, SMC, BMC, vinyl ester	3–5	110–200
6 Semi-aromatics, etc	PET, PBT, special PA, polyarylates	4–8	120–180
7 Aromatic polysulphones, etc	PSU, PES, PPS	9–14	150–220
8 Fluoropolymers	PTFE, ETCFE, FEP	5–20	150–260
9 Heterocyclics (including thermosets)	PAI, PEI, BMI, etc	15–70	170–250
10 Aromatic polyketones	PEEK, PEK	50–80	250–260

were included in Chapter 2. For detailed information, reference should be made to the polymer manufacturers or to the many databanks.

The versatility of engineering thermoplastics has been greatly enhanced by the development of blends and alloys. Naturally, this trend complicates the selection process, and makes simplistic judgements of material capability even less reliable. The properties and processing characteristics of alloys tend to be intermediate between those of the constituents, but good coupling technology can achieve a more useful profile than either constituent alone. Numerous examples are quoted in the application sections, but the most important varieties are these:

- Polyolefin rubbers blended with polypropylene to provide tough grades (e.g. PP/EPDM) for bumpers, instrument panels, etc.
- Polystyrene added to polyphenylene oxide (polyphenylene ether) to give a range of materials (mPPO or mPPE) with much improved processing at lower cost.
- PBT polyester added to polycarbonate to enhance its solvent resistance, making it suitable for bumpers.
- Blending of PC with ABS to give an optimum mix of impact strength, modulus and heat distortion resistance at an acceptable price.
- Dispersing PPO in PA 66, in order to combine the chemical resistance of the crystalline nylon with the creep and heat resistance of the amorphous PPO, for on-line paintable body panels.
- Rubber modification of PBT to make bumpers, and of glass reinforced PA to achieve seat structures.

The simplified display offered here recognizes that the first stage in the selection of a material must involve a process of elimination (see Table 3.1). A prime consideration for a motor industry material is its cost; another is likely to be the upper temperature limit for continuous service. In Table 3.1 therefore the ten principal polymer groupings are listed with guideline figures for cost (relative to basic PP) and continuous service temperature. There are many other basic eliminating factors, like chemical resistance, flammability and transparency, but these show too many differences within individual groups for their inclusion here to be helpful.

The first two categories in Table 3.1 account for some 80% of all automotive plastics, including as they do the three biggest volume materials, i.e. PP, PVC and PUR. Categories 3, 4, 5 and 6 – essentially the engineering plastics – make up most of the remaining 20% of applications. The sulphur-containing aromatics (group 7) and the fluoropolymers (group 8) are used in several crucial but small volume applications.

The last two categories (9 and 10) are scarcely seen at all in the motor industry, for reasons which will be evident from a glance at the cost column of Table 3.1. Nevertheless there are always exceptional applications to be found, where the choice of very expensive materials is justified by the special effects achieved. One example is the use of PEEK as a coating for shock absorber bearings; another is the use of sintered polyimide mouldings for unlubricated moving parts operating over a very wide temperature range in fuel management systems.

Materials selection

Traditional procedures

Different 'ad hoc' methods have been used in the past to establish plastics components in the automotive industry. Broadly, these methods conform to the following general sequence:

1. Define the needs.
2. Prepare a shortlist.
3. Select.

Stage 1 may well be 'taken as read', and the transition from 1 to 2 effected very rapidly, embracing only existing proven materials. The most care is likely to have been taken over stage 3, perhaps with a feasibility study with prototypes and a cost-appraisal.

Implicit in the preparation of a shortlist is a process of elimination. This is more likely to have been based on experience and general awareness, e.g., 'polystyrene is too brittle for a bumper', or 'PVC is too flexible for an instrument panel', or 'PES is too expensive for body panels', rather than anything more structured.

Experiments with unproven materials are discouraged in the motor industry, and until quite recently the quality of suppliers' data did not encourage a detailed selection exercise. Since the late 1970s two factors have combined to help make sophisticated selection procedures a reality. One is the generation of meaningful multi-point data by the material suppliers; the fruits of this were discussed in 'Designing with plastics' in Chapter 2. The other is the development of computerized databases.

Selection with databases

No attempt will be made here to assess different databases. They have been devised with differing objectives; the strength of some lies in the breadth of materials included, while others have strength in depth over a narrower front. All are involved in a continual process of evaluation and improvement. The need is to provide the designer with data which are directly comparable between one supplier and another, based on identical test methods and identical units. This insistence on comparability is the real fundamental change; the role of the computer is to transform the scope, speed and efficiency of the process.

What the database does essentially is to make it possible to achieve a meaningful shortlist. The method is to specify the upper and lower limits for the key properties, and ask the computer to select the materials falling within these bounds. However, two important points need to be watched. The first is that some key materials may be missing from the database. The second, more importantly, is that the material requirements fed to the computer may be inadequate. As always, the 'garbage in – garbage out' rule holds, and the more elegant the computer output, the harder it is to accept that the input may be flawed.

It may be said therefore that the key to successful selection is to define the needs correctly. Hence, whenever a new selection process is started, the items listed in 'The decisive properties' should be carefully addressed, and the upper and lower bounds should be defined, as far as possible. (In many cases this will be a matter of semi-quantitative 'ratings' rather than precise figures.) Mechanical properties such as flexural modulus, tensile strength and creep resistance were not defined as 'decisive' in the earlier section, because the overall rigidity, strength and deformation potential of components are a function of form and dimensions as well as basic properties. Nevertheless, feeding in the upper and lower bounds for mechanical properties is an important part of the selection procedure. The selection sequence can now be redefined thus:

1.   List the key properties.
2.   Identify the upper and lower limits of acceptance for each.
3.   Prepare a shortlist, using a database.
4.   Make final selection by testing.

The fourth stage, making the final choice from the shortlist, must involve practical testing, unless the materials and the application are very well understood. The reason is that invariably some aspect of behaviour will be missing from the database: it may have been forgotten, or more likely it is simply unquantifiable (like 'snap-fit-ability'). This stage may well merge with a typical motor industry feasibility study. Furthermore, sooner or later after production has started there is likely to be an optimization programme, or a cost reduction exercise. The value of a soundly-based shortlist, with a number of viable options, will then be confirmed.

Requirements for different application areas

This section summarizes the conditions which are encountered in each part of the car.

Body panels

Temperature:

In production:
Electrocoat oven	up to 200°C
Primer/sealer oven	up to 185°C
Enamel oven	up to 160°C
Repair oven	up to 140°C

In service in sunlight:
Horizontal panels	up to 90°C
Vertical panels	up to 75°C
Low temperature:	down to −40°C

Chemical agents:

Petrol, especially near the fuel inlet
Tar and salt, on the lower panels
Car polishes (usually abrasive)

Form stability must be retained at the upper temperature; impact strength must be retained at the lower temperature.

Any unpainted surfaces are expected to deteriorate no faster than the painted ones. Bumpers (i.e., front and rear ends) must retain impact strength and deformation resistance according to separate standards.

Exterior trim

The requirements for exterior trim are basically the same as for body panels. Some items will be fitted after the paint line, but may still be subjected to the repair ovens.

Transparent lenses for headlights and rear lights have to meet particular ageing specifications, with respect to retention of colour, impact strength, optical performance and scratch resistance.

Wheel trim designs have to satisfy requirements for impact strength, form stability (at disc brake temperatures) and paintability.

Interior trim

Temperatures: Below the windows   up to 80°C
                Windows and above   up to 90°C
                Fascia and parcel shelf up to 125°C

(In 40°C ambient temperature and direct sunlight, these local high temperatures under the front and rear screens really do occur, with modern low-angle 'aerodynamic' screens. Deformation problems are therefore more urgent with fascias and parcel shelves.)

Chemicals: Cleaning fluids and detergents: cosmetics;
             food and drink; sweat; spillages.

UV resistance is still relevant for interior parts.

Engine compartment

The harshest conditions of all are experienced in the engine compartment. In the average car however the total running time is typically only 5000 hours. The trends of low bonnets, reduced air flow and under-engine protection all increase the ambient temperature. The highest temperatures occur when the vehicle is standing after high-load use.

Temperatures:

Inside the engine:	
Normal	120°C
Peak: In the crankcase	up to 160°C
Under the rocker cover	up to 130°C
Coolant: Normal	110°C
Peak	up to 125°C
Within the transmission:	up to 150°C
Underbonnet generally:	up to 130°C
but locally	up to 250°C
Near silencers and exhaust cans:	up to 200°C
Near brake discs and exhaust manifold:	up to 350°C

Chemicals:

Engine and transmission:	Hot oil
Coolant:	Hot antifreeze
Fuel system:	Petrol

Brake and clutch controls: Hydraulic fluids
Battery: Sulphuric acid

Underbonnet components have to withstand spillages from any of the above chemicals, plus road salt, and survive impact from stone pecking, etc.

4

Interiors

General

The dominance of plastics

Inspection of any car model shows that the passenger compartment is dominated by plastics. Glazing apart, virtually every surface is a product of polymer chemistry, either as a solid surface or as a fabric. This is the area where plastics are most firmly established: in fact the passenger compartment accounts for some 56% of the total usage of automotive plastics.

The dominance however is less complete than it might appear. Table 4.1 shows that, even in the passenger compartment, plastics as yet have made very minor inroads into safety-critical structural and mechanical parts.

Comparison of modern high volume cars with those of earlier years, even up to the 1950s, demonstrates some surprising differences. Interior styling made few concessions to passenger comfort, or even passenger safety. In recent years, however, safety has been one of the major influences. Information gained from a grim catalogue of accident statistics is now backed up within the motor companies by massive programmes of sophisticated impact testing, with extensively monitored mannikins.

Historically, the most significant safety landmark for the passenger compartment was the American Federal Motor Vehicle Safety Standard FMVSS 201, the crucial feature of which was the head impact test on the instrument panel. More recently, occupant protection legislation has become much more demanding, and specifications such as FMVSS 208 present a wide selection of requirements such as the following:

- Head impact testing on instrument panel and pillar trims.
- Knee impact testing on certain areas of the instrument panel.
- Body impact testing on seat backrests.
- Side impact testing against door and side trims.
- Collapsibility of steering column.

66

Table 4.1. Interior plastics usage by function

Function	Share (%)
Rigid (decorative or semi-structural)	38
Foams	25
Flexible skins and carpets	20
Acoustic coatings	15
Structural and mechanical	2
	100

These safety considerations provide a framework, within which the needs (often conflicting) of fashion and function are played out.

Trends

Traditionally, the passenger compartment is the part of the vehicle which bears the highest assembly costs. The potential advantages of plastics in facilitating component consolidation and modular construction consequently make them particularly attractive in this area. Off-line assembly, complete with electrics, of sophisticated sub-assemblies is a logical extension of the benefits offered by plastics. This is now a well-established trend.

PVC was once the almost universal surface in car interiors. Although it survives as seals and gaiters, and as fascia coverings in combination with ABS and PUR, it has been largely displaced from seats and door panels. The current fashion is for textiles, mainly in nylon 66 and PET polyester. The trend extends not only to seating (where the 'comfort' benefits of materials such as warp-knit nylon 66 are obvious), but to semi-rigid door liners where the fabric covers a surface of resinated cellulose or polypropylene. For these applications polyester fabrics have been developed which can withstand the temperatures and pressures of injection moulding without pile crushing. Increasingly in Europe the fabric is applied as a trilaminate, in which the face fabric is backed by a layer of foam and a lining fabric.

Carpets are becoming more sophisticated, in line with the other surfaces, across the range of trim levels from needlefelts in nylon and polyester to more expensive tufted velours in nylon 66. In many cases simple carpet backing in PVC or low density polyethylene powder is giving way to more elaborate polyurethane foam mouldings which provide acoustic damping.

Aerodynamic styling trends are prescribing the almost universal use of severely sloping windscreens and rear windows. The additional temperature and UV effects are imposing very demanding modulus and heat distortion specifications on the materials for instrument panels and parcel shelves, and on the colour stability of plastic and textile surfaces.

Fashion and function

Without doubt, in today's competitive market the passenger compartment is a major selling feature of the vehicle. (This is just as true in the commercial sector of the industry.) One does not need to be a psychologist to acknowledge that the sales appeal registers on two levels – as 'cockpit' and as 'living space'. The cockpit demands a satisfying sophistication and complexity of functions, accessibly and attractively displayed; the living space must provide comfort and cossetting for driver and passengers. Figures 4.1 and 4.2, comparing two models from the popular end of the market, two decades apart, remind us how the requirements of both cockpit and living space have changed.

Every stylist knows that there are elements of both function and fashion in the design of every component. The great contribution of plastics has been to offer design freedom, enabling the stylist to cater for both of these elements. For interior surfaces, there are now a great many choices on offer:

- Hard or soft surfaces.
- Matt or gloss appearance.
- Smooth or textured finish.
- In-depth colour or painted surface.

Beyond these considerations, there is virtually complete freedom of shape and form. Above all, there is the possibility of integrating components, consolidating them into a better shape with an improved function, and all achieved with lower assembly costs.

For high volume production, the injection moulding process is the most effective in achieving low unit costs. The advantages of mass production can be combined with the needs of a fragmented, customized market by means of modular design systems, allowing several different assemblies to be produced from a few rationalized tools.

Plastic foams extend the stylist's design freedom enormously. Functionally, they are the basis of acoustic and vibrational insulation, and they can make a big contribution to passenger safety. Developments in polyurethane foam have been crucial to the improved designs and production efficiency in modern car seating.

The concept of sandwich assemblies opens up a very wide range of possibilities, using combinations of foams and textiles with solid skins and substrates. The following section looks at the specific characteristics and problems of plastics as surfaces, as distinct from plastics as structures.

4.2   Interior of popular small car (Ford Fiesta) from the 1990s (courtesy ICI).

4.1   Interior of popular small car (Vauxhall Chevette) from the 1970s (courtesy ICI).

Plastics surfaces

Any interior panel will have a structural or at least a supporting function as well as a cosmetic one. However, it is the surface which is the prime selling feature, and therefore a surface blemish must be taken seriously, even if it has no bearing on the operation of the panel.

Types of problem

Among potential problems with plastic surfaces are:

(All surfaces)
  – Light fastness.
  – Chemical resistance.
  – Staining.
  – Abrasion resistance.
  – Heat distortion.
  – Flammability.
  – Fogging.
(Hard surfaces)
  – Dimensional changes.
  – Noise.
  – Distortion.
(Soft surfaces)
  – Tear resistance.
  – Odour.

The common problems of abrasion resistance and of chemical resistance and staining have been effectively addressed in the materials now established in car interiors. The same can be said of flammability, as long as the present criteria based on rate of ignition and flame spread are considered adequate.

The increasingly severe windscreen angles are raising the performance requirements in terms of resistance to temperature and ultraviolet radiation. Materials for the instrument panel must now be able to withstand a static temperature of 125°C. The effect of increased UV activity is not quantifiable, but it implies a greater risk of colour instability and fading, and even mechanical deterioration.

Hard trim

The rigid surfaces most frequently encountered in car interiors are based on ABS and polypropylene. ABS has long been established in hard trim by virtue of its toughness, rigidity and scuff resistance, its reliable dimensional performance based on its low mould shrinkage, and its pleasing medium-glossy appearance. Polypropylene was first seen in interiors in the early 1960s in the form of contoured textured ('leather grain') trim panels. The distortion to which polypropylene was prone in large area or long mouldings could be disguised in a heavily contoured shape, and a uniform matt effect could be obtained in textured surfaces more easily than with ABS.

In recent years 'controlled rheology' (CR) grades of polypropylene, in which lower and more uniform shrinkage is achieved by narrowing the molecular weight distribution, have become widely available. These polymers, formulated across a wide range of stiffness/toughness characteristics, have greatly reduced distortion, the main technical limitation of polypropylene in this application. With matt-textured finishes very much in vogue, and with reported cost savings of up to 20%, polypropylene has displaced ABS from much of this sector. However, ABS is still preferred for many dimensionally-critical and appearance-sensitive parts, particularly in matt formulations. At a slightly higher price, PC/ABS blends offer a rather better combination of heat distortion resistance and dimensional stability than either ABS or PP.

Polyurethane provides a solution for low-volume production of rigid surfaces, in the form of reinforced RIM. A recent innovation is the use of glass platelets instead of fibres, giving rigid panels with superior surface finish. Structural RIM extends the versatility of PUR even further: lightly reinforced, it is suitable for semi-rigid panels, and with heavier reinforcement it can be used for key structures such as instrument panel armatures. Figure 4.3 shows an example of a 'high line' hard trim surface which is part textile and part PVC/ABS foil. The whole is backed in a single operation by inserting glass mat and producing a rigid PUR foam *in situ*.

4.3   Part-textile, part-plastic hard trim surface backed with structural RIM PUR, from Peugeot 605 (courtesy ICI Polyurethanes).

Soft trim

Leathercloth based on PVC paste was introduced as a seat covering on London buses as early as 1941. Morris Motors began to use PVC coated fabric for seats in 1944, and early volume-produced cars made very extensive use of PVC for nearly all interior surfaces. Compared with leather, the cost of materials and ease of production proved unanswerable arguments. Vinyls were never a popular choice for seat surfaces, however, and the development of improved knitted and woven nylon and polyester fabrics began the process of displacing PVC during the 1960s. The introduction of new variations of textured and perforated PVC sheet failed to arrest the change.

Apart from seating however, the vagaries of fashion still allow many options in the choice of surface, and some of these feature PVC. One solution which gives a bulky 'luxury' feel while retaining good abrasion resistance and good stretch performance is solid PVC on a backing fabric of non-woven polyester staple. For real energy absorbing properties, as distinct from 'soft feel', polyurethane flexible foam provides the best answer. Increasingly, the PUR foam is applied with the face fabric and lining in one operation, as a trilaminate with a foam layer up to 20 mm thick.

PVC is much too versatile to disappear from the scene altogether. PVC/ABS foil, backed by semi-rigid PUR foam, is still the most widely used surface for instrument panels. More recently, PVC has been blended with PUR to provide more temperature resistant surfaces. The desire to eliminate volatiles altogether from surfaces exposed to temperature is producing new answers: one of these is a flexible blend based on acrylonitrile, styrene and acrylic rubber. In 1993, Krauss-Maffei introduced a more radical development: a process for producing a sintered thermoplastic surface on a foam backing in a single operation.

Texture

Textured, grained finishes are popular both for hard and soft trim. (This is perhaps fortunate, because graining is an effective way of disguising defects and inconsistencies in a surface.) With PVC in its various combinations, the 'leather finish' is produced during calendering. The choice of grain can be more critical with injection moulded ABS and PP. Excessively fine graining is readily abraded, as are small 'pimples' resulting from a sand-blasted tool surface. Such damage produces a very obvious glossy effect in an otherwise uniformly matt surface.

Colour matching is easily affected by inadvertent smoothing of matt surfaces. Certain dark colours are particularly susceptible. Matching adjacent panels can be a very complicated procedure when different grades and different grain finishes are involved.

Fogging

Fogging is now a subject of major interest. Concern has arisen because of the emission of vapours of uncertain composition in the passenger domain, as well as the condensation of the less volatile vapours onto the windows. The problem has existed since nitrocellulose paints, leather finishes and volatile PVC plasticizers contributed to what was widely admired as the 'new car smell'. Today's new car smell can be less pleasant, but is undoubtedly healthier.

Fogging is associated with soft trim; the main ingredients are volatile PVC plasticizers (now being replaced by polymeric plasticizers) and products resulting from the interaction between PU and PVC. However lubricants and flame retardants have also been identified. The biggest area is the seat covers, but because of its temperature the instrument panel has been identified as the main source. The elastomers in mats, gaskets and gaiters also contribute.

Although modern car interiors experience an air change several times per minute, fogging is a very visible phenomenon and is regarded as a potential health hazard. The matter is now receiving widespread attention. In Germany there is an official test for fogging (DIN 75 201), and since 1987 there has been much investigatory work.

Plastics structures and panel applications

The sandwich concept

Traditionally, interior panels have been multi-material compositions, made up of two or more layers, involving fabrication processes with high labour content. One of the main thrusts of the industry in recent years has been to simplify and automate both the assembly operation itself and the fabrication of sub-assemblies for the passenger compartment.

This objective is being achieved by the use of many different materials and processes. As ever with plastics, the main theme is consolidation of components. For interior panels, this takes the form of integrating two or more layers in a sandwich construction. Examples include a textile skin formed over wood-filled polypropylene pressing, in a single operation, to produce a door panel, and a unit seat, in which the polyurethane foam is formed with the fabric facing already in position.

Materials and processes

The less severe the functional requirements, the more likely the material choice is to be determined by fabrication considerations. Nevertheless mechanical performance has some relevance in all interior components, and in several cases it is crucial. In general, it is the backing or support layer which determines whether the component meets the two limiting criteria of rigidity and impact strength.

It is useful to compare the requirements for different interior panels in terms of rigidity and impact strength. (This simplification ignores important characteristics like HDT and abrasion resistance.) Figure 4.4 is a schematic stiffness/toughness plot for PP formulations in the context of interior panels.

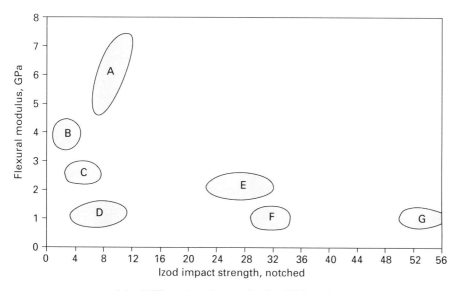

4.4   Stiffness/toughness plot for PP interior panel grades.

Table 4.2. Polypropylene compounds for interior panels

Class in Fig. 4.4	Formulation	Typical applications
A	Homopolymer + glass fibre	Seat pans, load floors
B	Homopolymer + 40% talc	Heater boxes, instrument backs, parcel tray supports
C	Copolymer + 20% talc	Consoles, door liners, A, C and D pillars, tailgate linings, steering column shrouds, quarter panels
D	Copolymer	B pillar covers, thermoformed bootliners, scuff plates, door panels
E	Copolymer, toughened and reinforced	Instrument panels, glove box doors
F	Heavy duty copolymers	Fascia underpanels, boot liners, wood-filled thermoformed panels
G	Copolymer/elastomer blends	Bumpers and spoilers

Table 4.2 defines these classes of formulation and lists the types of interior panel for which they are designed.

Where there is a separate surface layer, the choice of substrate is very wide. There are potentially cheaper backing surfaces, the surface finish of which is irrelevant, such as:

– Resin-impregnated fibre board.
– Wood-filled PP (pressed board).
– Glass mat reinforced PP.
– Corrugated paper (for roof linings).

The skin surfaces most commonly used in these two-layer systems are:

– Solid PVC, on knitted or woven fabric, or on needled polyester non-woven staple.
– Expanded PVC, either unsupported or on knitted fabric.
– Woven or knitted nylon or polyester fabric.
– Carpet fabric (for parcel shelves and supports).
– Thermoplastic polyolefin (principally in Japan).

With all these systems, a major consideration is the ease with which the two layers can be assembled in a single operation. Triple layer sandwich compositions can involve any of these skin and backing materials, with an intermediate layer of foam. A typical instrument panel sandwich would be made up of a skin of PVC/ABS copolymer foil over a semi-rigid PU foam. The backing or armature, formerly a steel pressing, is now likely to be injection moulded PP, or GMT/PP, or a resinated cellulose composition. The prime aim today is to eliminate one or even two of the assembly operations implied in any triple layer composition. The instrument panel/

fascia is a vitally important component and has probably received more attention than any other in this respect.

Instrument panels

The instrument panel embodies many of the conflicting roles of the modern vehicle. The range and function of the instruments and their presentation convey much of the sales appeal of the model, while the assembly contributes to the structural integrity of the vehicle and has to meet an increasing burden of safety requirements.

Throughout the 1950s, instrument panels in volume cars took the form of steel pressings painted in body colours. (The Volkswagen 'Beetle' and the Morris Minor were two typical examples.) Trends towards greater safety and complexity have rendered these designs obsolete. Originally the American FMVSS 201 head form impact test was the sole impact requirement; the specifications are now much more sophisticated, with a head impact test defined in Europe for passive passenger protection according to EG/ECE 21, and the new American legislation for automatic occupant protection according to FMVSS 208. The drive for plastic instrument panels also comes from the increasing number and diversity of instruments, the desire for 'soft feel' and leather-grained finish, for weight saving, and 'styling pressures'.

Plastic instrument panels first made their appearance in the USA in the mid-1960s, in the form of armatures in materials such as styrene-maleic anhydride copolymer. These were cost-effective, but with inadequate impact strength. (This class of material is making a comeback in new, tougher formulations.) The first single-layer plastic instrument panel was in polypropylene, for Autobianchi in 1974. This was quickly followed by the Fiat volume cars, concurrently with the first PP bumpers. Also in the 1970s, the Range Rover was given a single-layer instrument panel moulded in PP copolymer in one piece, and the carrier for the 'soft feel' panel in the Austin/Morris 1100 series was assembled from three mouldings in the same material. Some instrument panels subsequently appeared in polycarbonate (PC), but most of the running in the late 1970s and 1980s was made by modified polyphenylene oxide (mPPO). In the USA this was in laminated panels, but in Europe it was in single mouldings in high volume small cars such as the Ford Fiesta and VW Polo. By the mid-1980s, mPPO was beginning to be displaced by new 'second generation' PP formulations, improved both in stiffness and toughness, less resonant, and cheaper.

The PP-based formulations now have the largest share of the European market. They contain a carefully balanced mix of mineral reinforcement (usually talc, at between 20 and 40%), and rubber (usually EPDM). Glovebox outers, and other parts potentially in passenger contact, are similarly formulated. Resinated woodfibre pressings are still used as carriers, but the need to reinforce with some steel support makes this option less attractive than the newer plastics alternatives. Glass fibre reinforced ABS mouldings are used in several models, providing better dimensional stability than the PP systems. Another alternative is GMT/PP; this is more rigid than any injection moulded carrier, and can be cost effective in quite small run lengths.

Other current developments designed to create a more rigid carrier without weight increase are the use of blow mouldings or gas-assisted injection mouldings, or box sections made from injection mouldings. PP, ABS and PC/ABS blends are the likely materials.

A soft feel 'luxury' skin is still held to be essential for high-line passenger cars, at least for the upper part of the instrument panel. Much development work is in progress in the area of skin and foam, mostly directed towards making the manufacture of the composite panel less labour intensive.

The most widely used skin and foam combination is thermoformed ABS/

4.5   Fascia/instrument panel of Toyota Previa (courtesy GE Plastics).

PVC copolymer backed by semi-rigid PUR foam. This combination is more rigid than a 'soft feel' surface, but retains useful energy-absorbing properties. Two factors are threatening the status of this well-established system: the desire to simplify production, and the higher temperature specifications resulting from more aerodynamic windscreen designs.

An important 'simplifying' technique is slush moulding in PVC, first seen in Europe in the Audi Quattro in 1983. The essence of the process is that the PVC skin is held in the mould by vacuum while a semi-rigid PUR foam is produced behind it. In some designs glass mat is used as additional stiffening, in the form of structural RIM. Amine-free PUR systems are necessary to avoid the interactive effects which could occur with PVC/PUR composites at high temperatures. Because of its better temperature resistance, PUR is coming increasingly into the picture, even to the extent of a complete single polymer system in PUR, with an integrally-skinned foam, in-mould coated with PUR paint, and all supported on a base in structural RIM PUR.

The 1991 Toyota Previa introduced two new concepts in fascia design (see Fig. 4.5). The injection moulded fascia in mPPO is largely self-supporting, acquiring structural integrity from its deeply contoured shape (see Fig. 4.6). Furthermore, the moulding is given a soft feel and a less resonant surface by a thick coating of a filled PUR paint.

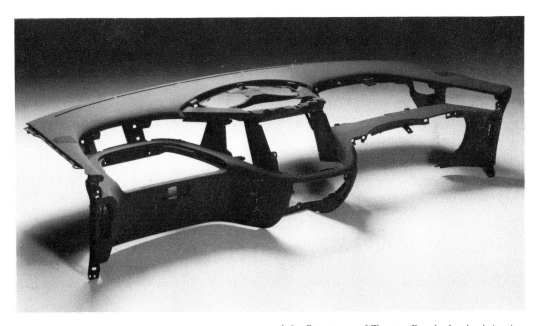

4.6 Structure of Toyota Previa fascia, injection moulded in modified PPO (courtesy GE Plastics).

Instrument panels present a changing picture, where the only certainty is further change. The trends towards cheaper assembly and higher temperatures are definite; the uncertainties are the 'fashion' factors such as texture and hardness.

Other impact-sensitive panels

None of the other panels forming the car interior is as complex and critical as the instrument panel; some are little more than space dividers, whose main characteristics are visual and tactile. Between these two limits there are several panels with clear strength requirements. Essentially they fall into two groups:

1. Panels where resilience and toughness are the main feature.
2. Panels where rigidity is important as well as impact strength.

These are represented by classes D and A respectively in Fig. 4.4. The first category includes parts which could impact the driver in a collision, such as 'B' pillar covers and the lower part of the instrument panel, and its adjacent trim panels. Unreinforced ABS and a variety of PP copolymers are commonly used for these components. The same tough, resilient materials are used for the components which need protection from the passengers' feet (seat valances, kick plates), or from their golf clubs (boot liners). Door pulls are also usually moulded from unmodified PP copolymers. Figure 4.7 shows a variety of interior panels in PP.

The second category comprises what are essentially structural parts, such as rear seat pans and rear seat backs which fold down to form part of the load floor. The most demanding application in this class is of course the load floor itself. Specifications for these parts are more likely to be based on flexural strength rather than impact strength. Glass fibre reinforcement is the common feature: injection moulded glass fibre reinforced PP is in use for seat pans, but load floors are more likely to demand the extra strength and deformation resistance of glass mat reinforcement. The matrix can be thermoset or thermoplastic; that is, unsaturated polyester or polypropylene.

Miscellaneous trim panels

The area where polypropylene has made most inroads into 'traditional' ABS applications is the very extensive general area of panels which require no great strength or resilience, nor high gloss appearance. The main needs are a well-reproduced grained surface, with good abrasion resistance and colour retention. Since the advent of controlled rheology grades, these attributes have been accompanied by good form stability and freedom from distortion.

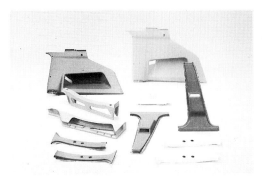

4.7 Interior panels for Rover cars, injection moulded in polypropylene (courtesy ICI).

4.8 Instrument bezel of Fiat Uno, injection moulded in ABS (courtesy Dow Plastics).

A typical grade will be a PP copolymer incorporating 30% of talc (class C in Fig. 4.4). Where extra rigidity is required, as in instrument binnacles or parcel tray supports, a 40% talc grade based on homopolymer may be used. Copolymers with 20% talc are suitable when more resilience is needed, as in glove box inners, side trim, and 'A' and 'C' pillar covers. ABS is still in use for many panels in this category, particularly consoles and quarter panels (see Fig. 4.8). Higher cost alternatives to ABS in high visibility panels such as pillar liners are blends such as ABS/PC and mPPO.

Textile finished panels continue to be in favour, especially for door panels and parcel trays. Resin-impregnated cellulose (chipboard) and wood-filled PP sheet are the most popular materials for these low-contour pressings, covered with woven or knitted polyester, and still, occasionally, expanded PVC.

Roof linings constitute a special case, with unique problems. The widely used basic design of a preformed sheet of resinated cardboard, bonded over the complete surface, is not really adequate for today's customers or today's production techniques. The trends are towards easier installation and a more luxurious, softer surface. Hence foam-backed knitted fabrics of various kinds are increasingly used, although there is still a significant market for expanded PVC. A newer low-cost solution is a moulding composition comprising textile fibre fleece bound with phenolic resin.

Acoustics

Nowadays noise is a very prominent consideration. In common with other 'creature comforts' like warmth, cushioning, and fresh air, the expectations of the travelling public in respect of quietness get more demanding year by year. Much has been achieved in reducing wind noise, road noise and engine noise; in consequence the minor squeaks and rattles emanating from the plastic parts constituting the interior trim are now much more noticeable. Small cars generally need firmer springing, in order to minimize the changes in handling due to the loading level; as a result they are more prone to vibration.

Improvements in primary acoustic performance, in terms of dampening of airborne and structure-borne sound, can be credited mainly to polyurethane. Flexible PUR foams are used for the insulation of the engine bulkhead, the floor pan, the roof, and the doors. Normally the flexible foam will be sandwiched between the structural panel and a 'massback', in the form of a heavy mineral-filled layer of PVC or polyolefin. Recent models, following the Fiat Tempra, feature PUR foam injected into the hollow sections of the body structure and the instrument panel.

The secondary acoustic characteristics (i.e., the squeaks and rattles) are attributed to movement between pairs of plastic components. The effects are sensitive to temperature and humidity, and are extremely difficult to track down. Lubrication and mixing of materials are generally helpful, and it is important that the trim fasteners retain sufficient torque with respect to temperature and fatigue. Variations in roughness and contact pressure are also relevant. PP is distinctly better then ABS over the entire frequency range, in terms of structure-borne sound damping, but in terms of airborne sound the difference is much smaller. Textile cladding has a small beneficial effect, but surface finish has a negligible influence.

The whole topic, because of its high customer awareness, is attracting a considerable amount of scientific investigation.

Structural and mechanical components

Seating

Seat structures are the area of car interiors where plastics have made the least inroads. Seat construction has long been recognized as one of the most labour intensive items in the production of a car, and so the motivation for drastic simplification of the assembly process is clear. However, the very high costs involved in designing and proving entirely new structures, in materials not generally accepted as 'structural', have been a powerful deterrent.

Several wholly plastic conceptual seats were presented in the early 1980s: notably one based on glass reinforced nylon for the Volvo Light Component Project, and another based on glass-reinforced polypropylene by ICI. These at least demonstrated the feasibility of plastic front seats. Another front seat prototype was developed jointly between Volkswagen and Bayer in nylon 6 reinforced with 25% glass fibre. A city bus seat in this material has been in use (to a simpler specification) by the Kiel GmbH company in Germany since 1982. No attempts were made to commercialize the ideas for passenger cars at that time. In a climate of increased safety awareness, attention switched to rear seats, which tend to present a rather less complex set of problems.

Rear seat backs and load floors had been developing in the USA for some years, mainly in GMT/PP. One of the first applications of this system in Europe was in the Mercedes T-series estate car in 1986. Here the GMT-PP system was selected after exhaustive comparisons with SMC and with steel and aluminium. Using a design involving two shells electromagnetically welded together, the plastic systems showed a 40% weight saving compared to steel and aluminium, together with significant cost savings. The device of splitting the rear seat asymmetrically is also relatively easier to achieve in a plastic structure. Within the plastics options, PP GMT was preferred to SMC on account of a lower weight and a less drastic failure mechanism. The advantages of glass mat stampings over short glass fibre injection mouldings in large area load floors are quite clear; the failure mode is one of progressive

4.9   Rear seat back for Audi 80 and 100, formed from GMT-PP (courtesy Elastogran/ BASF).

cracking rather than catastrophic disintegration. Figure 4.9 shows GMT-PP in an Audi seat back.

Traditionally, solid seat pans were regarded as inferior to frames with lattice springing. Improvements in polyurethane technology such as dual hardness foams have transformed the situation. Once the solid seat pan became acceptable, it was inevitable that steel should be replaced by one or other plastic material. The asymmetrical rear seat of the 1989 Peugeot 405 estate car features two pans in injection moulded glass-reinforced PP, with a variety of textile covering options (see Fig. 4.10). Other versions have followed on this theme, notably in PP-GMT.

The 1989 Ford Fiesta was the first example of a 'unit seat' in a volume car. The dual hardness foam was moulded directly in the textile cover, lined with a barrier film, with the fixtures already positioned for assembly within the ABS shell (see Fig. 4.11). The 1990 BMW 3-series featured bucket seats made in PP-GMT, in a process which included automated finishing and in-line painting, all within a one-minute cycle. These seats however still use steel for the safety-critical parts of the structure; the seat back and the seat sub-frame. In fact, new seat designs in production still owe more to develop-

4.10   Rear seat pan for Peugeot 405, injection moulded in glass reinforced polypropylene (courtesy ICI).

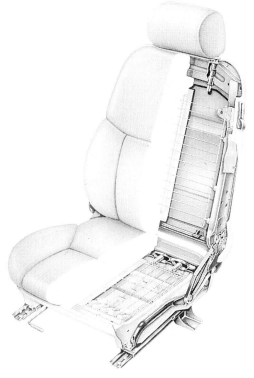

4.11   'Unit seat' of 1989 Ford Fiesta, including polyurethane foam-in-fabric with barrier film (courtesy ICI).

4.12   Seating in 1992 Renault Twingo: polyurethane foam-in-fabric without barrier film (courtesy ICI).

4.13   Typical heater assembly with housing in talc-filled polypropylene (courtesy ICI).

ments in PUR foam technology than structural design. The Renault Twingo seats, made by new no-barrier foam-in-fabric techniques, are a good example of this (see Fig. 4.12).

A truly plastic front seat was developed by Bayer between 1985 and 1989. The critical components – the back and sub-frame – are made in impact-modified 30% glass-reinforced nylon 6. This is a high strength material, but like all reinforced plastics it has a low elongation to break, implying fracture failure rather than deformation. This problem was overcome by the design of special shock-absorbers, made in a PC/ABS blend with very uniform characteristics over the relevant temperature range. The seat shell itself is in ABS. This concept seat survived the 50 km/hr rear impact test (which had failed many other plastic seats), and was reported to be fully compliant with the requirements of ECE R17, R21 and R25, and with the American FMVSS 201, 202 and 207.

The effectiveness of all-plastic structures has been conclusively demonstrated. There are many composites with the proven ability to perform as seat frames, backs and pans. However, questions of production engineering, cost effectiveness, and reliability still need to be addressed.

The increasing intrusion of recycling into the debate is highlighting the special status of S-RIM, in that it offers the possibility of a single polymer solution, covering both structure and filling.

Climate control

Forty years ago, facilities for the control of the environment within the passenger compartment of European cars were rudimentary, and for many basic models even a heater was an optional extra. Transformation to the

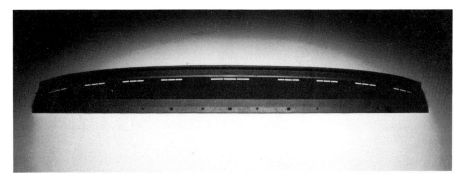

4.14   Demister grille for Volvo Trucks Belgium, in modified styrene-maleic anhydride (courtesy DSM Polymers).

present level of sophistication has of course been driven by competition and consumer demand, but greatly facilitated by plastics.

Since its appearance at the beginning of the 1960s, polypropylene has found use as a relatively cheap 'space divider'. Its ability to be produced in complicated shapes, by injection moulding or blow moulding, made it very suitable for heater ducting 'buried' behind the dashboard, being more compact and space-economical than corrugated PVC tubing.

Heaters and air conditioners are very similar structures. The largest component is the housing, and PP has emerged as the most cost-effective choice, usually as two talc-filled mouldings stapled or screwed together (see Fig. 4.13). Inside, the heat exchangers are aluminium with glass-reinforced nylon end tanks, very much in the style of cooling radiators. The blower wheels and fan are usually in acetal or glass-reinforced PP, and the controls which form part of the instrument panel assembly, in ABS. Steel (more reliably flat in thin sections!) is generally retained for the flaps, backed with PUR foam. Grilles for windscreen demisting, very long components requiring low shrinkage and good melt flow, are now being made in rubber modified SMA (see Fig. 4.14).

Door and window furniture

In earlier times interior door handles and window winders tended to be chrome plated zinc alloy die castings. Their size, shape and location could render them quite lethal in a collision situation, and considerable ingenuity has been exercised in designing more effective and efficient systems, with safer fittings at the passenger interface. Good use has been made of the performance capability and design freedom of engineering plastics (see Fig.

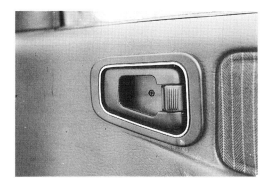

4.15   Interior door handle for 1982 Ford Sierra:
unfilled 66 nylon with surround in ABS
(courtesy ICI).

4.15). The handles are most frequently found in unfilled nylon, offering the optimum stiffness-toughness balance, but occasionally in acetal, to give better stick-slip characteristics. Door lock designs demand a mix of extreme rigidity and toughness; hence glass-reinforced nylon and some very flexible thermoplastic elastomers are involved as well as unfilled nylon and acetal. Electrically controlled central locking systems have created a demand for reliable low noise mechanisms.

Window designs have perhaps progressed even further. Remote controlled tape-driven systems provide a much more user-friendly product, using acetal for the window support and drive block and most of the moving parts. Motor housings are normally in glass-reinforced nylon. To convert the motor speed to an acceptably low speed of window movement involves a complex transmission mechanism. Special nylon-based products with internal lubricants have been developed for the necessary reduction gears. The window winder in Fig. 4.16 features a crank arm in glass fibre reinforced PET and a knob in acetal homopolymer.

Steering wheels

Traditionally the steering wheel has comprised a metal skeleton covered in plastic. For some 40 years the coating was cellulose acetate butyrate (CAB); hard, scratch resistant – and slippery in warm weather. Increasing safety awareness led to the dished hub, and to over-moulding in PP and PVC. More recently, in meeting the requirements of ECE 21 and FMVSS 203, improved performance opposite head and body impact requirements has been achieved by the use of integral-skinned PUR foam for wheel covers and hub padding (Fig. 4.17).

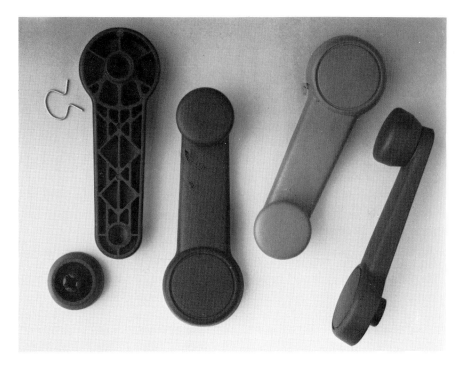

4.16   All-plastic window-winder in 1991 Ford Escort: acetal and PET polyester (courtesy Du Pont).

4.17   Steering wheel covered in integral-skinned polyurethane foam (courtesy ICI Polyurethanes).

A great deal of effort has been expended by material suppliers and motor manufacturers in designing steering wheels with a plastic structure. The difficulty is that the product needs to display high rigidity and good vibration resistance together with a ductile failure mode. It is unlikely that any one material can meet all these requirements. Hybrids may be the best answer, for example using a metal hub combined with reinforced nylon spokes and wheel. BASF have produced a hybrid composed of a rim made of filament wound reinforced vinyl ester and hub and spokes moulded from toughened reinforced nylon 6. Other apparently successful designs consist entirely of injection moulded nylon, either long fibre reinforced or toughened short fibre reinforced.

Clearly, both the cost and the safety factors will need to be examined in depth before plastic steering wheels are established in high volume car series. See Chapter 6 for details of plastics in the steering mechanism.

Air bags

Attitudes to the air bag principle and opinions as to its acceptability still vary from one market to another. Neither the designs nor the role of plastics in them have crystallized as yet. However, a brief review of the materials used in the first European airbag system, introduced into the Porsche 944 in 1987, is interesting:

- The air bag itself was woven nylon, with a flame resistant coating of chlorobutadiene. The gas inlet was reinforced with PBT polyester.
- The sensors comprised a beryllium copper spring and three polyetherimide mouldings assembled with epoxy adhesive.
- The cables linking the sensors with the diagnostics unit were sheathed in PVC, with end plugs moulded in nylon, PBT or nitrile rubber.
- The contact coils, translating the signal to the gas capsule, were housed in PC. The hub and gear tooth system of the coil was made of acetal and PTFE, with a centring device in an ABS graft polymer.
- The socket housing of the ignition pill was in glass reinforced nylon 6, with sacrificial caps in unfilled nylon 6.
- The cover ('deployment door') of the driver module was an S-RIM PUR moulding, incorporating a reinforcing nylon mesh and supported by an aramid fabric belt. The passenger module was similar but with a metal frame foamed into the cover. It was compatibilized into the dashboard with an aluminium sheet covered with PUR foam and ABS foil, which hinged open as the air bag inflated.

This materials assay was something of a tribute to the versatility of plastics. Subsequent designs have become simpler.

Seat belts

The advent of the air bag in no way supersedes the seat belt, which is still accepted as the most generally effective way of restraining the passenger, whatever the level of impact. Current seat belt designs are well established; the belts themselves are in woven nylon or polyester, and the inertia systems and fastenings are combinations of nylon, PET and acetal mouldings (but not excluding steel!), giving optimum rigidity, impact strength and frictional performance. More sophisticated retraction systems and height adjusters have led to even greater use of engineering thermoplastics.

Pedals

Plastics made a brief but successful entry into the area of control pedals in the early 1960s, when several General Motors models were equipped with floor-mounted accelerator pedals in polypropylene mouldings incorporating an integral hinge. These represented a two-thirds cost reduction compared with the steel and rubber equivalent.

More recently, cantilever designs of clutch and accelerator pedals in plastics have become established, particularly at Volvo and Saab. Glass reinforced nylon is the preferred class of material, the mouldings being gated so as to use the fibre orientation to maximize the strength of the component. The strength of the material is more than adequate, though it is however necessary to design around its low elongation to break by using a heavily ribbed section. Such a section in glass reinforced nylon will out-perform a steel U-section (which will tend to open out under flexural load), and also give a 30% weight reduction. Pedals of this kind were introduced by Ford in Europe in 1991.

An understandable caution prevails with regard to brake pedals. However, Ferrari have demonstrated that total reliability can be achieved in a plastics brake pedal. The 408 model in 1989 featured a brake pedal in glass reinforced nylon. The material was injection moulded nylon 66 incorporating 50% of long glass fibres (see Fig. 4.18). Fatigue loading over a range of temperatures at loads up to 900 kg failed to make any impression on these pedals after one million cycles.

Design and material selection are all-important factors in such critical components. Persuasive cost benefit and reliability figures are also needed before plastics can be generally accepted in these components in volume cars. It may be necessary to envisage complete sub-assemblies incorporating a reinforced plastics frame as well as the pedals. To this end, BASF reported in 1992 a pedal box in PP-GMT.

Instruments

Chapter 7 covers automotive electrics in more detail, but it is worth noting here that all the instruments to be seen in and around the instrument panel

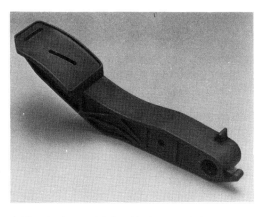

4.19   Instrument assembly: housing in PP, lens injection moulded in PMMA (courtesy ICI).

4.18   Brake pedal for 1989 Ferrari, injection moulded in long fibre reinforced nylon (courtesy ICI).

are themselves significant users of plastics. Non-reflective curved lens surfaces are PMMA mouldings, ventilation controls are mainly in ABS, and the switch housings are likely to be in nylon or PC. Membrane touch switches are now also beginning to appear, with membranes in polyester film. The instrument backs and housings are moulded in ABS or mineral filled PP. ABS has the advantage of ready assembly to the acrylic lenses by hot-plate welding, but PP offers cost and versatility benefits (see Fig. 4.19).

The automotive industry still caters for the smoker, although less generously than in the days of door-mounted and seat back-mounted ashtrays. New technology is represented by the polyimide linings now used in cigar liners. Ashtrays (here raised to the status of instruments!) have traditionally been made from phenolics. In recent years, however, flame retardant versions of nylon have been used in France and Spain, to give a front surface which could be colour matched to the rest of the instrument panel.

Other applications

Plastics are now involved in every working component in the car interior. Gear change mechanisms, handbrake controls and steering column switches and adjusters provide very good examples of 'hybrids', in which plastics and metals work together to maximum effect. PTFE is used (cost-effectively!) to prevent stick-slip problems in pedal hinge linkages and controls. The general theme is that the plastics provide the precision shapes and the rubbing surfaces, with steel rods or plates providing rigidity at low cost. All these applications are fertile ground for the engineering plastics: nylon, filled and unfilled, is the workhorse, followed by acetal and PBT.

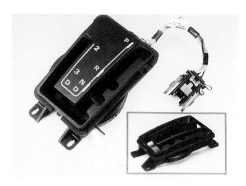

4.20   Jaguar gear selector plate injection
moulded in long fibre reinforced nylon; inset
shows moulding before assembly (courtesy ICI).

Many of the rapid assembly methods of today are dependent on ingeniously designed 'one trip' fasteners, most commonly moulded in nylon 66. Equally ubiquitous are the various trim clips in the same material, designed for rapid assembly and disassembly.

The functional range of plastics has been extended by long fibre reinforced nylon compounds. Jaguar introduced the first automotive application for this class of material in 1987, with a gear selector plate requiring great rigidity because of its large unsupported area (see Fig. 4.20). Further applications have followed, such as sun visor arms and control arms for external mirrors. These are minor components, but significant in that because of increased rigidity compared with conventional reinforced nylon, it has been possible to replace steel with cost and weight advantages.

5

Exteriors

Overview

History of car bodies

The original horseless carriages drew their design inspiration, naturally enough, from their immediate ancestors, the horse-drawn carriages. Strength and stability was provided by the chassis, and the coachwork was a separate entity. From the beginning it was made essentially from non-ferrous materials. This chassis-plus-coachwork concept was not seriously challenged until the 1920s. Had the matter rested there for another forty years, we might well have seen a smooth transition from timber frame and canvas to plastic car bodies. Traditional coach-building materials would have deferred to the increasingly clear design, cost and performance benefits of plastics.

Of course it did not happen like that. The beginnings of an integral steel structure were first seen in the early twenties with the Lancia Lambda; over the next fifteen years or so widespread use was made of unitized steel 'monocoque' structures, notably by Citroen and Opel. The huge post-war growth in automobile production was based almost entirely on unitized designs, and in recent years new technology has helped to make this system virtually unassailable for volume production.

The place for plastics

The dominance of the monocoque system derives from the consistency and relatively low cost of sheet steel, the facility with which spot welding can be automated, and the efficiency and sheer speed of press shop operations. Another consideration, often overlooked, is that sheet steel needs neither a chemical reaction nor a change of state in order to achieve its final shape.

These are all fundamental factors, which serve to contradict some of the over-enthusiastic predictions heard over the last two decades in favour of plastic car bodies for the volume market. In ignoring these realities, the plastics champions can do plastics a disservice. Set against big volume body shop practices, the manufacture, bonding and bolting of plastic panels are slow and expensive. But when they are set at the right targets, there is no doubt that plastics can be superbly successful.

The traditional bumper has now given way to complete integral front and rear ends, in which plastics feature very prominently. This is characteristic of all cars, whether large or small series. The same applies to the whole range of trim and 'hang-on' parts. With different weightings, the same benefits of plastics are exploited in all these applications. Weight saving, design freedom, component consolidation and corrosion resistance are axiomatic. Other benefits have been secured by research and development over many years: UV resistance, paintability, dimensional stability and impact strength are examples.

The 'numbers game'

The cost factors which determine the method of manufacture for any particular model are subject to specific local considerations, but there will always be a 'break even' volume below which the necessary investment in presses and press tools for a steel bodied vehicle is unjustified. It is below this qualifying level, which could be as low as 5000, or as high as 100 000, that some kind of chassis and space frame construction can be more cost effective than a unitized steel structure. Nowadays, opting for a space frame is likely to be followed by opting, at least in part, for plastic body panels.

Any comparison between different materials, regarding their suitability for an application, must be related to the processes involved and thence to the optimum levels of production. For something as large and complex as a car body, the analysis of the break even points is absolutely crucial. Factors such as model lifetime, production levels, tool costs, material prices and production rates must be taken into account, as well as less predictable market forces.

In descending order of break even points, the contending processes (i.e., potentially Class A finish) follow this sequence:

1. Steel pressing.
2. Injection moulding.
3. Sheet moulding.
4. Reaction injection moulding.
5. Resin transfer moulding.

(It would be unproductive to try to put figures to this sequence which would be applicable in every case, see Table 2.5.)

Several vehicle designs with wholly plastic panels mounted on a metal space-frame (or with some other steel reinforcing member) have been in production over the last forty years, with varying degrees of success. At the present time, the biggest sector comprises what can be described as hybrids, in which individual plastic panels are included in an essentially steel structure. These may be exploratory developments, or simply cost reduction exercises. More interestingly (and increasingly) they are part of a trend towards employing changeable plastic panels to achieve 'face lifts' and model variants.

Body panels and structures

Plastic cars

The first known plastic-bodied car was conceived by Henry Ford I as the 'soybean car'. A prototype was produced in 1941, with a tubular steel frame supporting 14 phenolic panels (derived in fact from petroleum, and not soya bean oil). Costs turned out to be far higher than anticipated (the panels were described as 'only a quarter of an inch thick'!). The car is best remembered now for the famous photograph of it resisting a sledgehammer vigorously wielded by Henry Ford (see Fig. 5.1). The Scarab of 1945 was a

5.1   Henry Ford I with the Ford 'soybean car' in 1941 (courtesy Ford Motor Company).

5.2   1974 Lotus Elite (courtesy Lotus
Engineering).

similar tubular steel framed prototype, made by Owens Corning to signal
the merits of glass fibre reinforced plastic panels.

Glass fibre panels achieved their first commercial success with the Chevrolet Corvette in 1953. This car was designed with unsupported glass fibre
reinforced polyester panels on a steel ladder chassis. With only very few
modifications, it has survived ever since, with production levels usually
exceeding 20 000 per year. SMC panels have long since replaced the primitive
hand lay-up system (see Fig. 2.6). The accumulated production total passed
the one million mark in 1992.

Lotus, producers of composite cars since 1957, took a big step forward in
1974 with its new Elite (see Fig. 5.2). This was manufactured by the new
VARI process (Vacuum Assisted Resin Injection), in which the two halves
of the body are moulded in matched tools, and subsequently bonded together
(see Fig. 5.3). The only chassis is a 'backbone' top-hat section in steel;
otherwise the stability of the body is ensured by integral foam beams (see
Fig. 5.4). These foam beams are used to locate and support uni-directional
glass rovings, before the polyester resin is injected. The result is an extremely strong body, of which the present Esprit is the latest example. The
Lotus system is extremely effective, but no claims are made that it will ever
meet the cost and production needs of the mass market. The De Lorean

5.3   The two halves of the Lotus body before bonding (courtesy Lotus Engineering).

5.4   Lotus construction: foam beams to locate the uni-directional fibre reinforcement (courtesy Lotus Engineering).

sports car project is remembered in Northern Ireland as an economic disaster. As a car body, however, it was spectacular and unique. The body was made by the Lotus VARI system, with reinforcing foam beams, and the styling was achieved by cladding the composite body with thin gauge stainless steel sheet.

A breakthrough was achieved in the 1960s by the Matra company in France. They used a process which they called 'Low Pressure Injection', now better known as Resin Transfer Moulding, to produce large panels for successive vehicles known as the Bagheera, Rancho and Murena. Progress in resin transfer moulding culminated, in 1986, with the Renault Espace, which has been one of the most successful examples of a steel framed, plastic bodied vehicle to date (see Fig. 5.5). The side panels are RTM, in-mould coated, weighing 28 kg, produced on a 5 minute cycle. Other panels use both RTM and rigid SMC; the problem of combining rigidity and impact strength in the doors has been resolved by mounting flexible SMC panels in steel frames. The additional rigidity needed by roof and bonnet panels is found by using a PUR foam filled core in a sandwich.

The second major landmark at General Motors, 30 years after the Corvette, was the development of the Pontiac Fiero. This used four different categories of material: SMC for the most rigid panels (the bonnet and the roof – a relatively small panel in this two-seater sports car), RIM and reinforced RIM for the side panels, and injection moulded PP-EPDM for the impact-resistant bumpers. The Fiero was an impressive indicator of the GM committment to plastics, and considerable sales were achieved between 1984 and 1988. However, the design, involving a rather heavy steel space frame, was less innovative than the assembly process, with its devices for locating the variety of plastic panels. The focus at GM moved in 1990 to the 'APV' (Advanced

5.5   Renault Espace.

Passenger Vehicle, described as a 'space transporter'). Designed to be produced in much larger numbers than the Renault Espace, the APV is the first high volume vehicle to feature large plastic panels. The large panels are in SMC, with the wings in R-RIM polyurea. Much of the body structure is composite, with only the underbody in galvanized steel. The APV could be described as the closest approach yet to an all-plastic bodied volume vehicle.

In the plastic bodied vehicles considered so far, the emphasis has been entirely on reinforced thermosetting materials; for the more recent models they have been formed by 'semi-industrial' processes, mainly SMC, R-RIM and RTM. The first significant use of thermoplastics (discounting some brave but unsuccessful attempts with ABS in the 1960s) was in the BMW Z1 roadster of 1987 (see Fig. 5.6). The different requirements of the individual body panels were reflected in the choice of materials:

- Horizontal panels: a complex multi-layered sandwich involving woven glass fabric in an epoxy RTM system, and inner layers of glass mat around a foamed PUR layer. This ensures high rigidity at low weight, with far better fracture performance than could be achieved with simpler (and cheaper!) systems such as SMC, plus excellent sound absorption.
- Vertical panels: injection mouldings in PC/PBT alloy, a resilient thermoplastic.
- Bumpers: a thermoplastic skin, injection moulded in modified PBT. The design as a whole passes the US 5 mph impact tests, largely because of the

5.6   1987 BMW Z1 (courtesy Dow Plastics).

supporting bracket and springs; composite structures made in RTM epoxy with carefully oriented long glass fibres.

The primary structure of the Z1 is galvanized steel, so in this respect the design of this 'plastic car' is orthodox. However, the body platform which forms part of this primary structure is plastic, and therefore broke new ground. Essentially this is made up of a PUR foam core, sandwiched within glass mat which is then impregnated with a foamed epoxy resin. The Z1 is bonded together by robots, but the cost of its specialized composite structures eliminates any possibility of this route being followed in a volume car.

Several other models appeared in the 1980s, continuing the general theme of wholly plastic skins on a steel space frame. The Reliant Scimitar has a body made up of panels in reinforced thermosetting resin; however, four different systems are used, illustrating how particular composites and processes are best suited to particular panels. Most are in polyester, with an RTM glass mat system used for the hood, where the greatest rigidity is needed, wet compression moulding for the rear deck lid and panel, and hand lay-up for the doors, underbody, wheel arches and boot inner panel. The panels needing greatest flexibility – the front and rear wings and bumpers – are in R-RIM PUR. A more specialized sports car, the Alfa Romeo SZ, was the first to be clad completely in double skin panels of thermosetting methacrylate resin, injected around glass mat in an RTM system. This polymer system cures more quickly than SMC, and by virtue of an in-mould acrylic gel coat, provides a Class A finish without additional steps. A similar construction was adopted for the Dodge Viper in 1991. Yet another variant in the field of specialist sports cars was afforded by the Treser, in Germany. This uses a space frame in aluminium instead of steel, to support a variety of plastic panels, mostly in R-RIM and RTM.

Hybrids

It is evident that the problems involved in designing a complete plastic car body can be quite daunting. However the growing competition for customers means that the lower tool costs and shorter lead times, together with un-doubted styling benefits, strengthen the interest in plastics. Virtually all the major motor manufacturers are examining plastic panel alternatives. At this stage in development, the usual answer is a hybrid, featuring partial material substitution.

The most familiar form of material substitution involves non-structural 'appearance' parts, such as wings (fenders). There are also several examples of semi-structural parts such as bonnets and tailgates, which have to support their own weight, and actually make a contribution to vehicle rigidity. More fundamental structural components are also being investigated, but inevitably most developments are concerned with panels which do not disturb the established assembly line and paint shop procedures. This simply reflects the enormous cost of laying down a new assembly line, or of making significant changes in an existing line.

There is a mixture of motives to be discerned in each plastic panel development. There have been some purely speculative adventures, charac-teristic of more affluent times; more commonly plastics development is designed as a weight-saving or production-simplifying step. The idea of using replaceable plastic panels to expedite model variations or mid-model face-lifts has been studied in Europe, particularly by Fiat and Volkswagen, but there has been no European equivalent of the American practice of making frequent changes of model appearance by means of fender extenders and front fascias in SMC.

In general the non-structural vertical panels are best served by unre-inforced thermoplastics. In fact, the first hybrid to include thermoplastic panels to any significant extent, the 1984 Honda Civic CR-X, used 11 kg of injection moulded PC/ABS alloy for front wings, side panels and front fascia, all painted off-line. A joint ICI/Austin Rover development in 1983 achieved a Class A on-line paintable polypropylene wing for the Maestro van, in the form of a sandwich comprising a glass fibre reinforced core and a toughened skin. Subsequently it became apparent that the sandwich solution

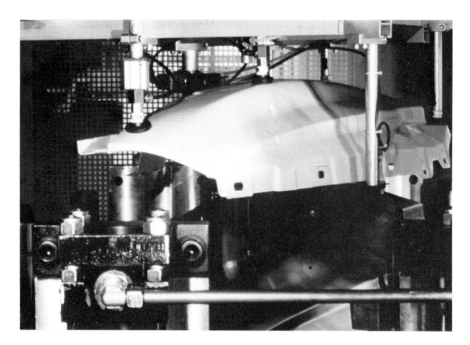

5.7 Injection moulded wing in PPO/PA from
GE Plastics for the Renault Clio (courtesy
Renault).

was unnecessarily expensive for a non-structural wing; the toughness, rigidity
and topcoat paintability requirements can be met directly by several thermoplastics, notably PPO/PA alloy, and possibly even by new formulations of
toughened filled PP.

In the Nissan Be-1 of 1986, PPO-PA alloy injection mouldings were used
for the front wings and front and rear aprons. They survived on-line painting
and oven temperatures of 150–160°C. Other manufacturers, in the USA,
Italy and France, have since adopted PPO/PA wings for small and medium
volume variants, e.g., the Renault Clio, shown in Fig. 5.7. This material is
claimed to combine the dimensional stability, low water absorption and flat
temperature curve of PPO with the chemical and temperature resistance and
good melt flow of nylon.

For more rigid panels, SMC is still the most widely used material. There
are several examples of well-established semi-structural SMC panels in
Europe. The best known is probably the bonnet of the Citroen BX; most
variants of this model have been using SMC since 1982. The Volvo 480 ES
now also has a bonnet in SMC. Audi have used SMC for the Quattro boot-
lid since 1983; now the tailgate of the Coupe is also in SMC. The first large-

5.8　Vauxhall Frontera Sport with hardtop
compression moulded in SMC by Autopress
Composites Ltd (courtesy DSM Resins).

area SMC part to appear in a high volume German car was the front end
(located just above the bumper) of the VW Passat in 1990. This is painted
on line and is an indication of the increasing industrialization of the SMC
shaping processes. Another new application is the hard top, with detachable
capping, of the Vauxhall Frontera Sport (see Fig. 5.8). The XMC modification of BMC has been used successfully for the Citroen BX tailgate since
the early 1980s, and BMC achieved high-volume acceptability with the Fiat
Tipo in 1990. A 1993 addition for BMC is the Citroen Xantia tailgate (see
Fig. 5.9).

RIM PUR is another body panel option. It has long been used as a
bumper material in small volume models, and in impact-sensitive panels
with a stiffness/toughness balance controlled by reinforcement in somewhat
larger series, notably the Pontiac Fiero. The Ford Fiesta XR2 features wheel
arches and rocker panels in RIM PUR (Fig. 5.10). Two newer developments
will keep the RIM concept at the forefront: polyurea, offering much faster
reaction times, and structural RIM, which uses glass mat reinforcement to
achieve much higher strength panels.

Resin transfer moulding, enjoying similar low cost tooling benefits and
potentially faster reaction times, is preferred for a variety of high rigidity

5.9 Tailgate of 1993 Citroen Xantia in BMC (courtesy Owens Corning).

5.10 Wheel arches and rocker panels in RIM-PUR (courtesy Dow Plastics).

panels. The Renault Espace, using an epoxy matrix, is the most conspicuous success to date.

Structures in plastics

Semi-structural components are becoming significant in supporting exterior panels, as well as underbonnet. Two contrasting examples are the front-end module of the 1990 Chevrolet Beretta in GMT-PP and the front end of the Citroen Xantia, in SMC (see Fig. 5.11).

There are many successful applications of this type, but nevertheless to move from these to a complete space frame would still be a big conceptual leap. The perceived wisdom at the present time is that any car body made up of individual panels must be based on a supporting space frame, and that the frame structure should be made of steel, or conceivably aluminium. Space frames in aluminium have in fact been used to support plastics panels in several concept cars, from the BL ECV-3 of the early 1980s to GE Plastics' Ethos II of 1993.

There have been many development projects directed towards the totally plastic body, based on a plastic space frame. A variety of composite structural materials has been used, including glass mat and carbon and aramid fibre reinforced compositions. To achieve the required rigidity, it has been necessary to design box sections, either from woven glass pre-pregs or a continuous fibre reinforced matrix, usually foam filled. Some extremely strong structures have been devised, but always at unacceptably high cost.

5.11   Front panel of 1993 Citroen Xantia in SMC (courtesy Owens Corning).

5.12   Underbody panel for Volvo 400 in reinforced polypropylene (courtesy ICI).

Furthermore, the bonding and bolting of panels to these structural members is slow and costly.

To achieve space frames in composites at acceptable cost, it will be necessary to design cheaper high performance materials, together with high speed automated production processes. Structures of high strength and low bulk would require something akin to a refinement of filament winding, for the main structural members. It has been suggested also that the RTM process, applied to knitted instead of woven preforms, could create 3D structures without discontinuities, probably to be used in conjunction with the main members.

Underbody panels

Underbody panels, or protection shields, may be classified as non-structural body panels. Their function is primarily aerodynamic and hence they are more likely to be found in high performance cars; however their role in protecting the engine compartment is useful. Such panels are not required to have a quality finish, nor are they subject to topcoat oven temperatures. The range of acceptable materials is therefore very much wider than for other body panels.

The most widely used material is GMT, in the form of PP-impregnated glass mat, hot stamped. Many American models have used this system since the early 1980s, using the benefits of low cost tooling and good production rates. However, where more accurate forming and location are needed, and when the numbers justify the higher tooling cost, some European manufacturers are turning to injection moulded panels. The Volvo 400 series of 1990, for example, uses a toughened glass fibre reinforced PP moulding compound (see Fig. 5.12).

SMC is not generally used for large area panels liable to be exposed to stone impact. It is however suitable for smaller, deep-drawn forms in this environment: the Audi spare wheel well in vinyl ester SMC is a good example.

The painting problem

Paints and plastics

In the days when plastics were first being considered for exterior panels, it was often said that these new materials foretold the end of the automotive paint shop, with all its problems. With plastics there was no corrosion and therefore no need for protection; unlimited pigmentation possibilities would provide in-depth solid colour across the complete spectrum, and high lustre surfaces could be achieved through high quality tooling.

These arguments have proved to be fallacious. Early adventures with ABS body panels exposed the poor weathering performance of the material, and the later vinyl foil-covered roof fashion proved to be short-lived. In the 1970s, the rapid fading of the black nylon mirror housings and SMC bumpers on some Renault models was taken as a bad advertisement for plastics. On the other hand, there are still several popular models fitted with 'sympathetic' pigmented bumpers rather than matching painted ones; and matt finishes are widely accepted as a styling feature on the lower parts of the body surface. However the market shows no sign of giving up its insistence on Class A finish for the upper parts of the body shell.

The two sectors of the chemical industry represented by paints and plastics were slow to realize their common interest in the automotive industry. It was late in the 1970s before a proper dialogue was visible between them. Subsequently there has been a fruitful and ongoing partnership, and some excellent results have been achieved. The problems arise on two levels. Firstly there are the questions of how to marry the properties of individual plastics to the practicalities of individual painting processes, and secondly (but for the motor industry, crucially) there is the question of how to accept these alien materials and systems within the constraints of the assembly line.

Problems of painting plastics can be grouped under four headings, dealt with in subsequent sections:

1. Temperature.
2. Adhesion.
3. Performance.
4. Appearance.

Temperature

This is the most important consideration of all. The hurdles which a potential plastic body panel must surmount (each with a dwell time of around half an hour) are these:

Electrocoat (primer-surfacer)	180 to 220°C
Topcoat	130 to 160°C
Touch up ovens	90 to 100°C
Off-line PUR systems	70 to 90°C

The ideal solution, for a hybrid, of passing the complete body-in-white through the entire process, is not really viable. None of the materials meeting acceptable cost criteria will undergo electrocoat conditions without some deterioration in form stability or in the paint finish. However, vinyl ester based SMC is reported as surviving the electrocoat cycle successfully, at least on low-visibility panels.

Normally, however, the plastic panels must be attached to the body as it passes down the line, immediately after the electrocoat stage. This is a disturbance to conventional assembly lines, but a tolerable one. Much less tolerable is the notion of breaking the line after the next stage, the topcoat baking ovens. For bumpers, this has been an accepted practice, at least for small and medium volume models, since the early 1980s; however for more basic body panels the risk of damage to the topcoated steel surfaces during assembly is considered too great. In any case, as the front and rear ends become progressively more complex and sophisticated, the traditional distinction between body panel and bumper becomes increasingly blurred. Hence the need for bumpers also to be topcoat-resistant is increasing, especially in high volume series.

There are many plastics options for body panels; quite the most critical parameter in the selection process is the topcoat oven temperature. Long term, there is great pressure on the paint suppliers to reduce the baking temperatures. In the short term, however, the pressure to convert to water based paints for environmental reasons is greater, and this works against any reduction in temperature.

Adhesion

Good adhesion is obviously desirable between a paint and its plastic substrate, although under extreme impact conditions, some loss of adhesion can be an advantage. The behaviour of different paint/plastics systems is very specific, and usually a slight solvent action can be beneficial. As always, a clean surface is necessary, and this can normally be secured by the usual degreasing procedures. The polyolefines used to be regarded as unpaintable, because of their unreactive surface, and special methods have been devised since the early 1980s to secure good adhesion. Flame treatment and cold gas plasma treatment are increasingly widely used; both enjoy the advantage of being solvent-free.

Performance

One of the advantages in using plastics for vulnerable panels like bumpers, spoilers and wings is their deformability and resilience; in fact their ability to sustain small impacts without visible change. For this advantage to be retained in a painted system, obviously the paint must exhibit the same deformation tolerance. If the paint layer is significantly more rigid than the substrate, the effect of an impact shock can be more serious than just a cracked paint film. If the adhesion is good, the crack in the film initiates a crack in the substrate. In fact, the ill-considered use of conventional topcoats can seriously impair the impact strength of resilient plastics. The problem can be alleviated by the use of flexible PUR primer coats.

Appearance

Class A finish is demanded by the motoring public. As long as steel and plastics co-exist in body shells, this standard is unlikely to be relaxed. Any new problems associated with the new materials are likely therefore to be greeted with some hostility.

SMC body panels used to be affected by 'outgassing' (the escape of volatiles during baking) leaving bubbles in the paint film. This problem has now been largely eliminated. Panels in PUR R-RIM tend to fall short of Class A finish because of surface roughness; this may be difficult to eliminate in a low pressure process. Reinforced thermoplastics experience a different problem: although the moulding can have an excellent surface, at paint oven temperatures the differential expansion between fibres and the matrix can result in an uneven surface on cooling. This can be hard to disguise without recourse to an additional paint layer. The best appearance is given by unreinforced thermoplastics; however very few candidates in this category can provide the necessary heat distortion resistance at an acceptable price. Some of the most successful thermoplastics body panels are alloys, in which the continuous phase is a high melting crystalline polymer with good chemical resistance, and the disperse phase is an amorphous polymer with a high transition temperature, e.g. PPO/PA alloys.

Summary

Compatibility with the assembly line will continue to be the dominant consideration. An assembly line that has been designed from scratch for a new steel/plastics hybrid is likely to be quite different from one which has been adapted to accommodate replacement plastic panels.

In the near future, off-line painting of plastic panels will continue. Several very successful two-pack PUR paint systems have been developed, and excellent matches to the topcoats have been achieved. Nevertheless it remains prudent styling practice to separate the two kinds of painted surface as far as possible. Long term, off-line painting will never be popular with the manufacturers (even for 'add on' front and rear ends) because of the dangers of mismatching and surface scratching, and of course the additional handling and cost.

In-mould coating should be mentioned here for completeness. This technique is widely used for integral skin foam components in PUR, principally in the passenger compartment. It is also used to prime SMC parts, as an aid to producing a Class A finish on the topcoat. The RTM methacrylic resin panels of the Alfa Romeo SZ were in-mould coated to a Class A finish. As this was a specialist vehicle and no other painted panels were involved, the usual problems of reduced production rate and poor colour matching did not apply.

Bumpers

History of development

Nowhere has the design potential of plastics been better expressed than in automotive bumpers. Significant restyling became evident in the early 1970s, and the sector attracted increasing development effort thereafter. The styling changes have been quite fundamental, but apart from looking better, the vehicles today are more aerodynamically efficient and show greater damage tolerance.

Minor exterior components first appeared in plastics during the 1960s. Foremost among these were radiator grilles. The grille is a high profile part, frequently used for marque delineation. Changing to plastics (usually ABS) not only effected a cost saving, but also allowed the design to be changed more frequently. Chromium plating of plastics, however, is expensive and not always satisfactory. A matt black finish was certainly a more logical option, and to some stylists, a better looking one. To what extent the change was inevitable, and to what extent it was contrived is uncertain; but the fact is that very rapidly in the early 1970s, 'black is beautiful' came to be the accepted wisdom, and the traditional insistence on bright chrome finish weakened very quickly. The way was now open for the plastic bumper.

One of the stages on the route was Ford's introduction of black rubber strips along the bumper in the 1973 Capri; another was the addition of plastic end caps to chrome bumpers, at Fiat and VW. The mix of black and chrome persisted as a styling feature for a while. The plastic end cap idea had obvious benefits: elimination of post-forming operations, the extension of 'wrap-around', with possibilities for styling variations, and conceivable benefits to pedestrians. Originally PUR was the selected material, particularly in Japan, with ABS and HDPE in use in Europe. However PP quickly took over everywhere, as UV stabilized copolymer grades. A common feature of low price cars for many years was the matt black bumper in rolled steel with PP wrap-around end caps.

A more fundamental innovation was represented by the Renault 5 in 1971. This was the first plastic bumper in Europe, a deep wrap-around design in SMC, painted grey; for a volume car, something of a leap in the dark. The benefits were functional and economic, but there was also a styling motivation. Deep side panels in SMC were added, to continue the line of the bumpers for the full length of the car and reduce the apparent height of this rather squat vehicle. These rigid front and rear ends retained their popularity in France for some years: the Renault 14 of 1977 also featured deep SMC bumpers, and the 1976 Chrysler Alpine was given similar bumpers in polyester RTM.

The main innovation was the introduction (starting with Fiat in 1975) of a complete bumper cover, usually in EPDM/PP, to give damage resistance at up to 4 km/hr impact, according to ECE 404. In most cases a reduced bumper beam was retained for the greater part of the width of the car; in other cases the cover was supported only by fixing plates. In 1978–9 most of the Fiat and VW range adopted designs of the former type, and the latter appeared in Poland on the Polonez. These bumpers were a straightforward replacement of bright chrome by matt black plastic, but they also provided greater damage tolerance (reputedly a source of instant popularity with drivers of Fiat taxis). As black or dark grey injection moulded EPDM modified PP these bumpers survived in many model ranges, including the GM Cavalier/Ascona range, until the late 1980s.

Renault apart, the concept of the complete plastic front or rear end was slow to develop in volume cars. Lotus used PUR foam-filled ABS in the mid-1970s, while British Leyland used PUR and various thermoplastics in a range of small-volume sports cars. Their Triumph TR7 in 1976 was designed with bumpers in injection moulded PC, although the rear bumper was converted to glass reinforced nylon 12 when the fuel sensitivity of polycarbonate became apparent. What was to prove the main trend for the future

5.13   Front bumper of 1979 Citroen Visa, in
polypropylene/elastomer blend (courtesy ICI).

took off in 1979 when the Citroen Visa and Fiat Ritmo went into volume
production with EPDM/PP bumpers which fitted into the overall body
contour and actually replaced some body steel (see Fig. 5.13).

The 'watershed', when for the first time there were more models being
manufactured with partly or wholly plastic bumpers than with wholly steel
ones probably occurred in 1979, which was also the year when the 4 km/hr
legislation was introduced. From this point on, the 'ultimate' ambitions for
bumpers began to take shape: the idea of a completely integrated front and
rear end, suitably colour matched, and totally damage resistant at 4 km/hr.

5.14   1984 Rover 200, with painted
polypropylene bumper (courtesy ICI).

5.15   1993 Ford Mondeo, with bumper in PC/
PBT blend (courtesy GE Plastics).

5.16   Rover Montego, with bumper from Bayer
in PBT/elastomer blend.

In the early 1980s, the 'hard' front end, usually in SMC or other forms of glass reinforced polyester, gave way to the 'soft' front end. Polypropylene was by far the most widely used material, usually blended with an elastomer such as EPDM. One landmark was the 1984 Rover 200 (see Fig. 5.14), the first series model to use painted PP, employing a PP formulation and paint priming system developed by ICI. There have been two important exceptions to the PP tendency. A blend of PC with PBT was introduced by GE Plastics for the 1982 Ford Sierra; this is stronger and more rigid than PP, and hence more self-supporting in deep, wrap-around sections than PP. The cost disadvantage has been largely offset by the introduction of easy flow grades, permitting thinner sections. This trend has been carried through successfully to the Mondeo of 1993 (see Fig. 5.15). The 1983 Maestro from Rover featured a rubber modified PBT composition from Bayer; unlike blends based on PP or PC, the PBT-based materials can be topcoat painted on-line. This feature has continued in subsequent Rover Maestro and Montego models (see Fig. 5.16).

Modern requirements

The degree of protection afforded by a bumper is defined in a variety of national and international regulations and in-house specifications. The common feature of most specifications is the 4 km/hr impact resistance; the details vary in terms of the pendulum and its directions and contact points. Testing at 8 km/hr, a much more severe condition, is increasingly requested, and this requirement has a major influence on bumper design for models intended for export from Europe into the USA.

Bumpers need to be resistant to deformation as well as fracture. Some of the early deeply contoured designs with resilient plastics were in fact too elastic. They failed to absorb sufficient energy, and transmitted too much to the main body structure, without registering any permanent damage themselves. The importance of a balanced design is now appreciated, and three components are recognized as being necessary:

1. The skin, with appropriate appearance and abrasion characteristics.
2. The energy absorbing medium. Originally this would simply be ribbing, or a partial box section, or even a blow moulded section, but PUR foam is now most commonly used.
3. The armature. This is the main protection for the body shell. Steel is usually preferred, but the weight saving potential of aluminium and SMC has led to many successful designs using these materials.

In choosing the skin material, one decisive factor is the major styling question of whether the front and rear ends are to be painted in body colour. Things are clearly easier without the need to achieve integrated colour. Without it, the material can be selected on factors like impact strength, dimensional stability, and resistance to abrasion and UV. With the need for colour matching comes either (a) the problem of the end sections being assembled after the topcoat stoving stage, and achieving a perfect match between two different paint systems, or (b) the problem of selecting a material which is not only form stable at 140°C or more, but also remains tough after coating with a brittle topcoat. A flexible PUR primer may be necessary to ensure this.

Typically, the central part of the skin is made into a box section by welding on a back plate in the same material. The assembly process can be simplified by creating the foam layer between these two surfaces. The skin section is usually attached to the armature by brackets specially designed to allow for differential thermal expansion. All the attachment mechanisms, including those between the armature and the chassis, have some bearing on the performance of the complete assembly.

More expensive cars tend to have extremely elaborate bumper designs. Two-colour systems are increasingly favoured (following the Opel Omega of 1987); some Mercedes models use a superficial layer of soft PUR foam outside the main energy absorbing mass of foam, with an additional abrasion strip outside the main bumper skin. Elaborate designs of brackets have been conceived, incorporating a spring function as well as a 'dashpot' function, in order to control the energy absorption process with more precision, and achieve 8 km/hr impact performance.

Variations will persist, depending as always on model specification, volume of build and assembly plant circumstances. Certainly what used to be bumpers, and are now integrated front and rear ends, are major styling features. To the two main design considerations, styling and energy absorption, can now be added a third – recyclability. This is already beginning to affect design and material selection, because complex heterogeneous bumpers are not easily recycled. The problem is discussed in Chapter 8.

Other exterior components

General

As with the interior, the exterior components must meet the needs of both function and fashion. A functional part like a door handle needs to have a good appearance (and retain it for the life of the car), and a largely decorative part such as a lateral strip clearly has a protective function, as well as a secondary styling function of masking an interface. It is accepted practice to assemble most of these components after the topcoat oven stage; nevertheless a degree of temperature resistance is necessary because a significant proportion of all finished vehicles will be exposed to the touch-up ovens, at around 100°C. Designs are still being developed for lateral protection strips, in the interests of easier assembly and reducing the risk to the newly painted body surface. The 1992 Opel Astra uses an innovative system, in the form of a contoured plasticized PVC section co-extruded with a base section in glass-reinforced PVC, which needs only adhesive to be fixed to the body.

Chromium plating of plastics components virtually died during the 'black is beautiful' styling revolution. However, the popularity of bright finish has been sustained in some up-market vehicles, and there are signs of bright finish re-entering the styling scene. The loss of toughness accompanying plating is now better understood, so the technique is not applied to flexible components. On the other hand, there can be functional advantages, such as UV stability, EMI shielding and conductivity. ABS is still the most suitable plastic substrate for plating. A bright finish can also be obtained by second surface metallizing, by deposition of aluminium vapour on the rear surface of transparent materials, but this technique is not much used nowadays beyond decorative motifs and bezels.

Grilles

The radiator grille was the first significant exterior component to be converted to plastics in volume cars, in models such as the Renault R6 in 1968 and R12 in 1969. ABS, with and without reinforcement, has remained the prime choice for this application, often with grades providing a matt surface and improved low-temperature impact strength (see Fig. 5.17). ASA is sometimes preferred, for its UV resistance, while reinforced nylon, PPE, PP and PBT have all been used in cases where the temperature conditions are more severe. The attraction of plastics for the application derived from the versatility and cost advantage of the injection moulding process. Latterly the desire to improve aerodynamic performance has threatened to eliminate the radiator grille altogether.

The radiator grille is not about to disappear, however, because it is frequently a high profile feature, designed to be as conspicuous as possible in marques such as Mercedes and Volvo. The extractor grille, on the other hand, is designed to be inconspicuous. Frequently it is built into the 'C' pillar: a design of this type, on the Ford Mark III Cortina of 1977, was one of the first painted SMC mouldings in Europe. SMC is still used in this application, e.g., on the Citroen BX. Glass reinforced nylon, injection moulded and painted, was used for the grilles mounted on the 'C' pillars of

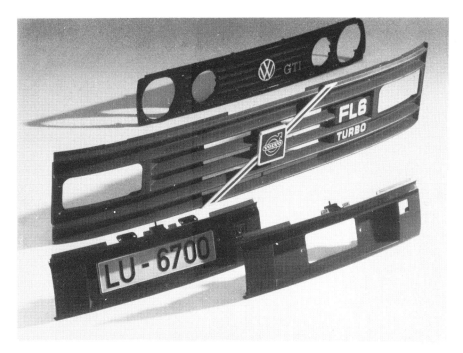

5.17 Mouldings in ASA: front grilles and rear
trim panels (courtesy BASF).

5.18 Cowl panel grille in PC/ABS blend
(courtesy Dow Plastics).

the Mercedes 190 and Volvo 300 series, for example, and for the grilles located on the rear wings of the 1988 Nissan Bluebird.

The cowl panel grille, between bonnet and windscreen, can be a more demanding application. It needs to be resistant to weather and hot air, but more difficult to achieve in this very long, thin moulding are good mould filling and dimensional stability. Glass reinforced nylon has been used, but amorphous plastics offer lower shrinkage and thermal expansion and so are more suitable. ASA and ABS are widely used, or PC/ABS blends where greater heat resistance is needed (see Fig. 5.18).

Similar properties are required in the rear trim (applique) panels between the tail lighting, usually featuring ABS and ASA (Fig. 5.17). An integrated styling effect has also been achieved by using an acrylic moulding to link the two acrylic lenses.

Spoilers

Rear spoilers tend to be the hallmark of the high performance top-of-the-range models, wherein aerodynamic refinements can be claimed. Production numbers are therefore relatively small, which explains why R-RIM PUR is the most commonly used material for rear spoilers, either produced in black or painted off-line in body colour, using very flexible PUR paints. Front spoilers are exposed to high wind forces, and consequently the material needs rigidity as well as toughness (see Fig. 5.19). Spoilers are increasingly becoming standard items, needed in much larger numbers. Injection moulded mineral-filled PA and PBT are used, along with blends such as PC/PBT, as well as SMC and R-RIM PUR. The problem of combining high rigidity and high impact strength in one component has been solved in the case of the 1993 Renault Safrane by using a thermoplastic elastomer for the impact-vulnerable spoiler base.

Mirrors

The door-mounted rear view mirror with a plastic body, now a virtually universal feature, was a French innovation, introduced with the Renault 14 in 1976. The great majority of today's mirror housings are moulded in glass reinforced nylon, in a variety of formulations based on both PA 6 and PA 66. Black UV-stabilized grades are normally used, with body colours an added refinement for high series models (see Fig. 5.20). Reinforced nylon is used because of the mechanical performance requirements of remote-controlled mirrors, especially the creep resistance needed for the mechanisms to survive the touch-up ovens. The precise grade selection depends on whether die-castings are used for any of the structural elements (still the

5.19 Front spoiler of Ferrari F40 in a polyurethane integral skin system (courtesy Dow Plastics).

5.20 Ford mirror housing in glass fibre reinforced nylon (courtesy ICI).

case with most designs). Bayer have taken the development to its logical conclusion by designing a mirror housing in nylon 66 reinforced with 33% glass fibre, which excludes metal entirely.

Materials other than nylon are increasingly entering this field. Plastics from BASF are used in the novel double mirror design for a Kaess-Bohrer Setra bus. The foot housing and the housing for both mirrors are injection moulded in ASA/PC blend, and the whole assembly is encased in vacuum-formed ABS.

Door handles

Exterior door handles need the same kind of properties as mirror housings. Glass reinforced nylon is popular in Europe; however glass reinforced PBT and polyarylamide are used for certain Ford, Peugeot and Volvo models, and acetal homopolymer for some truck cab handles (see Fig. 5.21). Initial

5.21 Iveco truck cab handle in acetal homopolymer (courtesy Du Pont).

fears about the wisdom of making access handles out of potentially inflammable materials have been largely overcome. Edge-free surfaces in these materials are in fact very difficult to ignite; nevertheless, some countries do specify flame retardant grades.

Wheel trim

For very many years, the after-market has supplied items intended to customize the wheel faces of mass-produced cars. The modern full-face wheel trim is much more than this. It derived from the drive for fuel conservation of the late 1970s, followed by the application of aerodynamic principles to car design, and the popularizing of the 'drag coefficient'. Certainly, smooth full-face wheel trim can make a contribution to drag coefficient at high speeds, by reducing the turbulence in the wheel area. Additionally, vanes are designed in to control the cooling of the wheels, and the basic need remains to keep dirt out of the wheel bearings. Few would doubt, however, that the main influence of wheel trim has been in marketing rather than in aerodynamics, simply by identifying the level of trim in a series, and by signalling the model year.

Small diameter hub caps in nylon had been used by VW, Ford and GM in the 1970s, but the first full-face trim appeared with the launch of the Ford Sierra in 1982 (see Fig. 5.22). With two different designs of trim, the principle of delineating high and low line models by their wheel trim was established at the very outset. The wheel trim habit is now all but universal in Europe, and shows no sign of subsiding. One result has been to reduce the sales of alloy wheels; these are still used when the extra cost is justified in terms of reduced unsprung weight facilitating improved performance, but are much less used simply for styling.

The application has proved to be a rich one for engineering plastics. The requirements are quite severe. Creep resistance and dimensional stability are necessary for the covers to stay in position in use over a wide temperature range, especially when disc brakes are generating heat. At the same time the covers need to be flexible and resilient enough to withstand kerb impact and stone pecking and the rigours of repeated installation and removal. Furthermore they need to be painted to the standards of a high profile body part.

Although ABS and PP in special formulations are now seeing some use in cost saving measures, this application has been dominated throughout by nylon, both with and without reinforcement. The reasons are the temperature resistance and the overall stiffness-toughness balance, and the comparative ease with which formulations can be devised to meet the needs of individual designs, often involving different fixing systems. These are based on both PA 6 and PA 66, with a variety of additives. Some formulations

5.22   1982 Ford Sierra: an early full-face wheel
trim design in glass fibre and mineral filled nylon
(courtesy ICI).

extend to no less than four prime ingredients: nylon, rubber, glass fibre, and
mineral particles, for example.

Road wheels

The concept of dramatically improving performance by reducing unsprung
weight by means of plastic wheels has exercised the motor industry's in-
novators for many years. The arithmetic – 10 to 15% lighter than aluminium
and 30 to 45% lighter than steel – is indeed attractive. The requirements
however are extremely demanding: the two extremes to be reconciled are
impact resistance (simulating high speed kerb contact) and creep resistance
(simulating the temperatures of wheel hubs and brake discs at their limit of
performance).

Successful plastic wheels have been produced, in the context of sports
vehicles. Wheels moulded from polycarbonate were in use for rallying in the
former USSR and other Eastern Bloc countries in the 1970s, based on a
Bulgarian process for injection moulding very thick sections without voids.
Wheels of this type however would probably not pass modern specifications

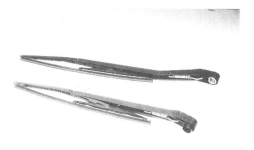

5.23   Windscreen wiper arm on fibre reinforced
PET (courtesy Du Pont).

for high temperature creep. In the USA the Motor Wheel Corporation has
successfully proved sports wheels compression moulded from special SMC
and BMC compounds based on vinyl ester, with high strength being achieved
by careful locating of continuous glass fibre. Doubts have been expressed
however about the inability of these wheels to deform on impact and so
avoid piercing the tyre wall. Developments are continuing, but there is little
sign of the cost benefits that might stimulate the necessary work on material
formulation and wheel design.

Sun-roof components

Although concealed for most of the life of the vehicle, sun-roof frames and
guiding systems do tend to emerge when the sun is shining. Hence they can
qualify as exterior parts. Glass reinforced nylon has been widely employed
in this area, but the 1991 Ford Escort used glass fibre reinforced PET for
what was the world's first all-plastic guiding system. Maximum rigidity and
dimensional stability were the motivation for this choice.

Windscreen wiper assemblies

Gears for windscreen wiper motors are remembered as one of the first (and
most enduring) examples of the exploitation of nylon as the 'workhorse'
engineering polymer, back in the 1950s. Wiper arms, expensive to fabricate
in steel, are nowadays widely produced in reinforced PET and PBT, which
combine high rigidity with UV resistance and rapid production capability
(see Fig. 5.23).

6

Engine, power train and chassis

The engine compartment

Whether the culture describes it as 'under the hood' or 'under the bonnet', and whatever the climatic extremes, the engine compartment is essentially a difficult environment for plastics. Unavoidably, polymers based on the carbon-carbon link are heat-sensitive materials, but nevertheless engines run more efficiently at high temperatures. This apparent mismatch is continually being made worse by design trends, which are inserting more components into smaller spaces with more restricted air flow. (Many of these trends are responding to legislation, in respect of fuel consumption, emissions and noise abatement.) Attack from a variety of chemicals at high temperature and bombardment by mud and stones complete the picture of a hostile environment.

The engine compartment of course encloses most of the vehicle's sophisticated precision components, traditionally designed around the tight dimensional tolerances of metals engineering. Furthermore, the automotive industry has developed a culture of reliability and cost effectiveness, in which high risks and adventures are definitely not encouraged. In view of all these factors, we can take it that any plastics application which has become established in the engine compartment is there by right!

Over the last forty years or so there has been a steady increase in the variety and scope of underbonnet fluid containers made in plastics. Washer fluid reservoirs in LDPE were the first; much more demanding applications have followed, with HDPE and PP as cooling system expansion tanks and nylon for hot oil containers. A spectacular recent arrival from BASF is a 50-litre hydraulic oil reservoir for trucks, blow moulded in nylon 6 and with very high molecular weight.

Reference to the 'Why plastics?' question addressed in Chapter 1 suggests that in the engine compartment the main driving forces for these successful applications have been reduction in corrosion, noise and maintenance requirements, together with the ever-present potential for consolidation of parts and easier assembly. Some of the many success stories are described in this chapter.

However, along the way there have been many cases where plastics applications have not blossomed into production, in spite of proven material performance. The reasons are complex, but usually involve the difficulty of justifying the very high costs of developing and proving changes in critical machinery. At the present time, worries about recycling and possible attendant legislation are having an additional braking effect on the whole process of metal replacement, in spite of potential long term ecological benefits.

New high performance polymers and composites are still emerging from the material suppliers. Often they achieve some success in aerospace or defence applications, or even on a Grand Prix Formula One 'test bed'; almost invariably, however, they fail to clear the motor industry's 'cost barrier'. This situation looks even less likely to change in the 1990s than it did in the 1980s. In 1988 BMW attempted to quantify the cost qualification for engine components: it was suggested that, above 150°C, each extension of the high temperature capability by 10°C was worth about 2.6 DM/kg. However this figure might be manipulated or updated, the inescapable conclusion is that the most expensive high performance plastics (such as the last two categories in Table 3.1) will never achieve a significant presence in volume cars. The motor industry (and its customers) will always find a way of managing without them.

The cooling system

Air movement

Plastic cooling fans and fan shrouds first came into general use in the early 1960s. There were obvious advantages in replacing laboriously fabricated sheet metal parts by one-piece injection mouldings. The recently discovered PP, offering good mouldability and properties at low cost, was the clear favourite for fans. Higher ambient temperatures, or greater air movement, or a higher creep resistance requirement at the hub have often necessitated moves to higher specification materials. An early landmark was the introduction of a five-bladed nylon fan by Mercedes in 1968. Other similar innovations followed (see Fig. 6.1). Higher temperature alone would suggest a switch to glass fibre reinforced PP; higher strength specifications would suggest glass reinforced nylon, whereas high impact strength requirements would lead to an unreinforced nylon. For climates with a high dust and grit incidence, Peugeot have introduced fans in long-fibre reinforced nylon 66. The market in the 1990s has been entirely captured by PA and PP, both

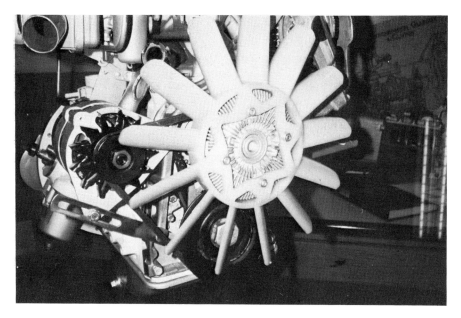

6.1   Early British Leyland cooling fan in
unreinforced nylon.

6.2   General Motors fan assembly, with fan in
reinforced polypropylene and support in mineral
filled nylon (courtesy ICI).

reinforced and unfilled. Figure 6.2 shows a typical fan assembly: the fan is in glass reinforced PP and the supporting 'spider' in mineral filled nylon 66.

For fan shrouds the main requirement is for rigidity, often in quite thin sections, rather than impact strength. Glass reinforced materials are therefore almost universally used, especially reinforced PP, with SMC and GMT-PP increasingly evident as well as reinforced nylon. In the 1980s Peugeot pioneered the development of complete front panels in which the fan shroud is consolidated into a larger structure.

Coolant circulation

The major success story for plastics in the cooling system is that of the radiator end tank, or 'water box'. This was first devised in France in the early 1970s by Sofica (now Valeo), for the Volkswagen Passat. It replaced the traditional labour-intensive and somewhat leak-prone copper-brass radiator by a solderless aluminium assembly, in which the plastic end tanks were attached to the radiator cores by direct crimping. This was a clear innovation; significantly, neither Sofica nor VW were directly involved in water-based engine cooling systems at that time.

Exhaustive testing in hot antifreeze under pressure established glass fibre reinforced nylon 66 as the best material for the end tanks. Twenty years and many millions of tanks later, it has not been necessary to modify this conclusion, beyond minor formulation changes to improve hydrolysis resistance (see Fig. 6.3). Some French and Italian models used radiator tanks in mPPO at first, because of the dimensional advantages conferred by a low-shrinkage amorphous material, but in the long term the fatigue resistance proved to be inferior to reinforced nylon.

The dimensional problems encountered in designing a leak-proof assembly marrying aluminium to a relatively high shrinkage anisotropic material can be severe. Two main factors contributed to the solution: firstly, employing substantial, well-designed gaskets to accommodate the dimensional differences between the two materials, and secondly, allowing for the anisotropy of the material in designing the injection tool. Expertise in tool design evolved with time, in such ways as deliberately maximizing anisotropy in

6.3   Selection of Valeo radiators from 1970s, with end tanks in glass reinforced nylon (courtesy ICI).

6.4   Rover expansion tank in reinforced polypropylene (courtesy ICI).

order to produce straight-sided mouldings, and compensating the tool in cases where curvature in the moulding could not be eliminated.

Engine coolant is a very exacting medium: the glycol content can exceed 50%, and the temperature can peak at over 130°C, with a pressure approaching 2 bar. It is perhaps surprising that an inherently hydrolysis-susceptible polymer like nylon 66 can perform so well; that it does so is a tribute to the strength of the bonding between the glass and the polymer. The other factors are the relatively short working life of a car (typically 5000 hours), and the fact that only the inner surface is in contact with aggressive chemicals.

Water pump housings (and thermostat housings) have long been targeted by plastics suppliers. Dimensionally these are very demanding items; engineering phenolics seem to be the most likely contender.

Nowadays, expansion tanks are also in plastic materials. Typically, these

6.5   Expansion tank integrated with radiator
end tank, in a single moulding in glass
reinforced nylon, produced by Blackstone for
Citroen (courtesy ICI).

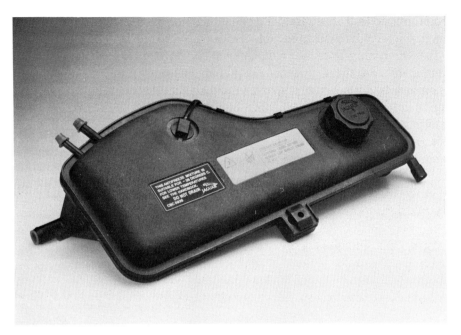

6.6   Expansion tank in reinforced nylon from
the Jaguar XJ6 (courtesy Du Pont).

are blow moulded in either PP or HDPE, or assembled from two injection
moulded halves in glass reinforced PP by hot-plate welding (see Fig. 6.4).
The trend to consolidation has seen the appearance of systems where the
expansion tank is joined directly to one of the end tanks, or even injection
moulded together with the end tank in one integrated moulding (see Fig.
6.5). A different solution in the high performance Jaguar XJ6 is a separate
expansion tank in glass reinforced nylon (see Fig. 6.6).

There are further innovations which simplify assembly by consolidation.
Cooling water pipes are produced in both PP and PA, with the flanges and
fixing lugs being moulded in, together with the necessary pipe bends. Both
thermostat housings and water manifolds have been established in glass
reinforced PA 66, and recently these two mouldings have been integrated
into a single moulding.

Underbonnet structures

Front end frames

These various largely two-dimensional structures, now constituting an important group of plastic components, originated with the Peugeot VERA

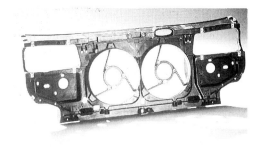

6.7   8 kg front end frame for Peugeot 405
moulded in SMC by Inoplast (courtesy DSM
Resins).

concept car of 1983. By 1987 Peugeot (working with Owens Corning Fiberglas) had established three different systems for manufacturing fan support panels, which were a logical development from simple fan shrouds, into larger (2 kg or so) structures supporting several functional parts, and enclosing the front end of the body-in-white. The 205 model had a frame injection moulded in nylon 66 containing 20% glass reinforcement, replacing a fabrication of 27 metal parts (see Fig. 1.4). For the 309 model a similar and significantly heavier part was produced by hot-flow stamping a 40% GMT-PP of the 'structural thermoplastic composite' (STC) type from Exxon Chemical. The Citroen GSA, Visa and CX were equipped with extended fan shrouds moulded in SMC. The same principle was involved in all three versions; the choice was based on the usual mix of strength and creep requirements, fixing method, likely processing wastage, numbers required and initial investment. Unsurprisingly, different models resulted in different choices.

The largest structures of this type are those developed for the Peugeot 405 and 605, and Citroen XM. These are single pressings in SMC weighing 8 kg (see Fig. 6.7). New industrialized processes for SMC, coupled with robotized placement of inserts and assembly, have resulted in the efficient production of a large structure to which thirty other components are attached. The concept has been taken up by other manufacturers, and several materials and processes are under investigation. The 1992 VW Golf (at around 1.3 million units annually representing virtually the ultimate in volume production) uses GMT-PP for this internal front end (see Fig. 6.8). At 3.7 kg it is smaller than the big SMC components, but it fulfils sixteen different functions.

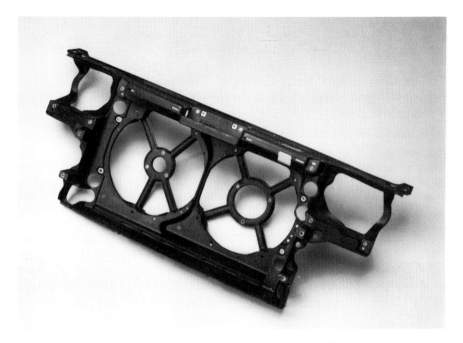

6.8   Front end frame for VW Golf formed from
GMT polypropylene (courtesy Elastogran/
BASF).

Heavier structures

Attempts have been made to harness the enormous strength potential of composites to make more basic structural members. Engine brackets in long glass reinforced polyester have been evaluated in VW Golf models: the strength was found to be more than adequate, with a weight saving of 3 or 4 kg. However there were doubts about the torsional rigidity, and the design was clearly not cost effective. Daimler Benz experimented with engine brackets in epoxy, reinforced separately with aramid fibre and carbon fibre: these survived 50 km/hr collisions, but still displayed a slight cost disadvantage. Ford have evaluated engine cross-members in RTM, using Dow vinyl ester as the matrix.

Further experiments can be expected using placement technology, with high concentrations of uni-directional fibre used in RTM, RIM or SMC moulding systems to provide very high strength in specific locations. The potential weight saving and performance advantages are likely to encourage more development projects along these lines, although the chances of an early breakthrough into volume car production are remote.

Transmission

Bearing cages

Transmission systems contain enormous numbers of plastic components whose existence is unsuspected by the motoring public. Daimler Benz have stated that a modern car may contain over a hundred plastic parts within its transmission system, totalling no more than 1 kg in weight. Bearing cages are among the most numerous and the most ubiquitous of these. Plastics have been used for injection moulded roller bearing cages since the early 1960s, and now feature in crucial assemblies such as the clutch release, differential, universal, gearbox and road wheels.

The most commonly used material is nylon 66. Heat stabilized unfilled nylon can be used in long term temperature conditions up to around 90°C, providing the loads are not excessive. Glass reinforced PA 66 is used with continuous service temperatures up to 120°C, and whenever the creep loading is likely to be too high for the unreinforced polymer. Above a 120°C working temperature, the long term oxidation resistance of PA 66 is suspect; in this high temperature area (up to 180°C continuous), PES has become established, usually in the glass reinforced form (see Fig. 6.9). (In fact, above 90°C, PES is much superior to bronze.) Newer, more modestly priced high performance polymers like glass fibre reinforced PA 46 are now competing for the middle ground of 120 to 180°C working temperatures. Nylon 46 is in use as gearbox supports, working in a similar environment (see Fig. 6.10).

Clutch components

This is another area containing many small, largely unnoticed, but very effective and cost-saving plastic components. Nylon has been an important

6.9 Gearbox bearing cages in polyether-sulphone (courtesy ICI).

6.10   Gearbox supports in nylon 46 (courtesy
DSM Polymers).

ingredient for many years, in such items as thrust washers, clutch release
bearings, clutch rings and master cylinders. PA 66 is the most widely used,
with PA 46 appearing recently in high ambient temperature applications.
Engineering phenolics are being used in reactors for automatic transmissions,
offering lower costs than aluminium and better creep resistance than
thermoplastics. Clutch facings feature several high performance materials,
with aramid fibre (also used in automatic clutch V-belts) often replacing
asbestos. Reinforced PTFE is established in this area, and PEEK has been
successfully evaluated.

Clutch hydraulics have been dependent on plastics for many years. The
fluid reservoirs are most commonly made from injection moulded PP, but

injection moulded PA is also used, with and without reinforcement. Blow moulded containers in PA 6 have recently made their appearance.

Gear shift mechanisms

This is a fruitful area for specially developed formulations which combine lubricity with good creep and temperature resistance. The Audi 80 gear shift mechanism is composed of two snap-fitted components in a nylon-based composite containing PTFE and silicone as well as glass fibre; in another case an actuator tube in reinforced PPS moves in contact with the PEI housing for an automatic gear shift. (Both these examples feature formulations from the LNP Company.)

A related application is gearbox covers: these were made in glass reinforced PA 66 for the Renault 5 as long ago as the late 1970s.

Propellor shafts

The potential advantages of continuous fibre reinforced composites in high specification 'uni-directional' parts such as propellor shafts have attracted designers for some years. There are considerable benefits in reduced noise and vibration to be gained, as well as the obvious weight reduction. Successful propellor shaft designs were developed for small volume production in the 1980s by VW and Renault. A very comprehensive investigation was carried out for the Audi Quattro Sport in 1990: the propellor shafts developed in carbon fibre reinforced epoxy performed extremely well under abusive conditions, including crash testing. Costs, however, were still far beyond the levels of volume cars. In 1993 production of the four-wheel drive Quadra version of the Renault Espace began, with a 1.6 m filament wound drive shaft. The reinforcing fibres were 75% carbon and 25% glass, with an epoxy matrix.

Engine 'hang-on' parts

Air filters

Polypropylene became accepted during the 1970s as an effective and low cost 'space divider' for the air filter housing, exposed to heat and chemicals directly above the engine block. Various formulations have been used, with glass reinforcement or mineral filling being incorporated for the larger designs. Injection moulding ensured cost savings against sheet metal fabrications. In most air filter assemblies, the filter itself is held together by a high density PUR foam, which also forms an airtight gasket between body and cover. Air inlet tubes, often located close to the exhaust manifold, are normally moulded in glass reinforced nylon, because of its greater heat deformation resistance.

Oil circulation

Components which are in contact with oil in and around the engine are expected to survive at least 5000 hours, at temperatures of around 130°C in normal running, and occasionally touching 160°C. For parts outside the crankcase the running temperatures are more likely to be in the 110–130°C range. Oil filler caps are normally in heat stabilized nylon 66, and PP has been used for larger caps incorporating a breather. Oil filter housings are made in reinforced nylon and engineering phenolics.

Oil pan (sump)

This application has attracted much attention from the plastics suppliers, without (it seems) commensurate reward for their efforts. The main difficulties are the temperature (where an occasional 160°C is realistic) and the tendency for leakage between the plastic and metal surfaces. There is also the fact that most sumps are relatively simple steel pressings, and not too expensive to fabricate. In general, only when the design involves a very deep draw or multiple pressing is there likely to be a clear cost benefit in changing to plastics. Evidence from oil and cooling water temperature measurements indicates that changing to a composite sump would allow the engine to reach its optimum operating temperature more quickly; but this alone is insufficient as a cost incentive.

Unreinforced plastics are insufficiently rigid, and indeed insufficiently strong, if the possibility of the car being jacked up via the sump has to be entertained. By the same token, short fibre reinforced mouldings are less likely to be acceptable than long fibre or glass mat reinforced compositions. Nylon 66 reinforced with random glass mat was accepted for a time in the USA in the early 1980s, but was soon withdrawn. Versions of SMC with vinyl ester as matrix have been proven for some heavy vehicle sumps, and for the Porsche 944 in 1987. Competition here was from aluminium; the SMC oil pan showed a one-third saving in both weight and cost, and survived the necessary test for impact strength and flammability.

Rocker covers

Plastic rocker box covers have been on the horizon for many years. There was a false dawn in the USA in 1981, when AMC introduced a cover moulded in glass fibre and mineral filled nylon 66. Like many other concepts, this failed because of leakage (or rather, because of failure to make design changes to accommodate the differential expansion between the cylinder block and the rocker cover). Subsequently, redesigning the flange and increasing the number of bolts, and using a gap-filling silicone gasket has enabled filled nylon to be used successfully in the USA. Nylon is used in Europe for the Citroen AX, in the guise of a filled polyarylamide (semi-aromatic PA), but VW selected reinforced PET for its VR-6 engine. However, because of dimensional stability and uncertainty about behaviour under extreme overheating, polyester-based BMC is preferred by several manufacturers, including Ford. Vinyl ester-based SMC is used for some large diesel engine covers in the USA.

Replacement of metal rocker box covers has progressed so far much less than was anticipated in the early 1980s. The probable reason is that the cost advantage can easily be swallowed up in the design modifications which necessarily accompany the change. The most effective replacements are made when it is possible to incorporate some extra component in the cover moulding, and so simplify the overall assembly operation. The valve cover of the 3 litre V6 engine as fitted in the popular Ford Taurus is made of vinyl ester based BMC, and incorporates integrated design features and a specially designed gasket. In addition to reduced noise and vibration, a weight saving of some 40% is achieved.

Engine covers

Many models include an engine cover nowadays; its function is partly cosmetic, recognizing the relevance in the saleroom of a good-looking engine compartment. There is also a practical function, i.e., to protect some of the smaller components, particularly the electrical circuitry, from dirt and

6.11   Mercedes engine cover in glass fibre reinforced nylon (courtesy ICI).

damage. The mouldings may be merely space-dividers, but the hostile environment ensures the use of reinforced PA or PP. Mineral filled nylon is widely used in the USA. A Mercedes engine cover in glass fibre reinforced PA 66 is shown in Fig. 6.11.

Intake manifolds

One of the most important development objectives for plastics in the under-bonnet area throughout much of the 1980s and 1990s has been the intake manifold, especially for fuel injection engines. Here there are clear proven advantages for plastics: in cost saving, and in up to 50% weight saving, and in improved performance on account of the smoother internal surfaces. Furthermore, lower thermal conductivity leads to quicker engine warm-up, hotter running, and lower hydrocarbon emissions.

6.12   Intake manifold for BMW in glass reinforced nylon (courtesy BASF).

The problems involved in making intake manifolds by injection moulding (discussed in Chapter 2) meant that the process development and production engineering entailed have been extensive and expensive. The preferred material has emerged as PA 66 reinforced with 35% glass fibres. The first successful production for a volume car began in 1990 with the BMW 6-cylinder engines, in BASF material (see Fig. 6.12).

Thermoset materials have been used in an essentially similar fusible core method, notably in BMC for the Ford Escort 1.8 diesel. Ford found, however, that reinforced nylon provided better resistance to vibration, fatigue and impact than BMC, and in 1992 adopted 35% glass reinforced PA 66 for its Zeta engine. The 1.8 and 2.0 litre versions use a Bayer grade, while the 1.6 litre engine uses a similar material from Du Pont. (This last, shown in Fig. 6.13, was the first intake manifold incorporating exhaust gas recirculation.)

The system was extended to V8 BMW engines in 1992, in mouldings weighing 3.5 kg (see Fig. 6.14). This must rate as one of the most complex

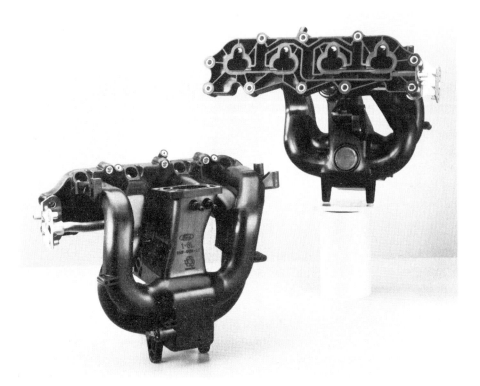

6.13 Intake manifold for Ford Zeta 1.6 litre engine, with exhaust gas recirculation (courtesy Du Pont).

6.14   Intake manifold for 1992 BMW V8
engine, involving 90 kg of fusible cores (courtesy
BASF).

achievements to date in fusible core technology. The core comprises ten
separate components, and when assembled weighs over 90 kg.

Rover have reported the successful production and road testing of an
intake manifold made in Du Pont glass reinforced nylon simply by injection
moulding two halves and vibration welding them together. This eliminates
the weighty problem of handling the metal cores, but can only be employed
when the matching surfaces are geometrically simple and when the internal
operating pressure is not too high. BASF announced in 1992 that their glass
reinforced nylon is being used in this way for Peugeot diesel engines (see
Fig. 6.15).

6.15   Intake manifold in glass reinforced nylon
for Peugeot diesel engine, made by welding two
mouldings together (courtesy BASF).

Underbonnet fuel components

Variations in the composition of fuel around the world are beginning to
have an effect on the choice of plastic material for these components. Nylon
in its various forms has excellent resistance to petrol. PA 66 and 6 are still
satisfactory in 85/15 petrol/ethanol fuels, but are unacceptable with fuels
containing 15% methanol. For such fuels, acetal (POM) is preferred, or (if
the cost is bearable) the more expensive PPS or PES.

Fuel piping (traditionally copper) is now almost universally nylon 11 or
12, having proved itself in service for over 30 years. It is much easier to

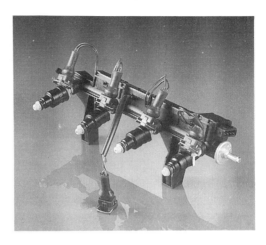

6.16   Fuel rail for VW in glass fibre reinforced nylon (courtesy ICI).

install, being either flexible or 'semi-rigid' and readily heat-formed, and is sufficiently resistant to methanol. However, future fuel emission standards, especially in the USA, and the increasing use of alternative fuels, will make life more difficult for PA 12. Fuel lines in the USA are beginning to take the form of multi-layers, such as co-extrudates of PA 12 with ETFE (ethylene tetra fluoroethylene) fluoropolymer. In-line petrol filters are normally nylon 6 or 66, i.e., both the monofilament mesh and the moulded housing. Carburettor floats, once nylon, have been made for many years in acetal. After a long history of unsuccessful attempts the carburettor itself has been produced, in Japan, as an injection moulding in engineering phenolic.

For fuel injection engines, the expensively fabricated metal fuel rails are increasingly being injection moulded in glass reinforced nylon 66 (see Fig. 6.16). The change has provided faster production, more compact installations, and better performance. Ingenious quick-acting connectors have been designed in nylon for use with fuel rails.

Modern fuel management systems require extreme resistance to wear, friction and fuel, and also require dimensional stability. Bosch are using sintered polyimide from Du Pont as spacers, thrustwashers and bushings for fuel pumps, throttle adjusting motors and mono-jetronic idle control units.

Emission control

Emission control legislation is providing challenges and opportunities for plastics. Emission control canisters first appeared as heat stabilized unfilled

PA 66, in the USA in 1969. Nowadays glass reinforced versions are more commonly used. PA 46 has been selected for the housings of the valves controlling exhaust gas recycling and secondary air supply; these applications involve ambient temperatures of 140°C, and which can peak at up to 200°C. These housings also have a snap-fitting capability on assembly, and good fatigue resistance.

Miscellaneous

There are many other plastic components functioning close to the engine block. Nylon 66 has long been used for speedometer gears, and more recently for cruise control parts. A major new application area for plastics is noise shields, for engines and gearboxes. The shields are relatively simple shapings, and can be produced without expensive tooling. GMT-PP is most used, but SMC and RTM have also been reported in Germany. This application is legislation-driven, and will therefore become even more widespread.

Space restrictions and the need to protect individual components from 'hot spots' are generating a market for specialist fasteners, usually in unfilled nylon. Figure 6.17 is an interesting example of such a multi-functional component.

Turbochargers are a new area of application for plastics; the hoses in particular have to meet very arduous conditions. Together with 'hot spots' and restricted space, there are the additional problems of low temperature impact, vibration fatigue and resistance to oil mist. For the 1991 Rover M-16 engine, the turbocharger hose was made in an ethylene-acrylic elastomer, extruded around fabric reinforcement in an oval profile (see Fig. 6.18).

6.17   Fuel filter holder in nylon 66, manufactured by FSM, Poland, for the Fiat Cinquecento (courtesy Du Pont).

6.18   Turbocharger hose for Rover M-16
engine in fabric reinforced ethylene-acrylic
elastomer (courtesy Du Pont).

Engine interiors

Timing mechanisms

Timing belt pulleys and timing chain sprockets are produced in PA 66, but engineering phenolics are now competing for this application. Timing belt covers are made in reinforced or mineral filled PA 66, or in filled PP where the temperature conditions permit. Timing chain tensioners require good creep and abrasion resistance at high temperature; the needs have been met by nylon variants for many years. PA 66 filled with molybdenum disulphide has been very successful, but is now being challenged by PA 46, which is better suited to these extreme conditions than PA 66 (see Fig. 6.19).

An innovative development concerned with engine management of twin camshaft engines has been introduced by Brampton Renold in France: a key feature is the use of PES for the linear-to-rotary piston. The requirements are the ability to be moulded to close tolerances, with good creep and abrasion resistance in the presence of hot lubricants.

6.19   Timing chain tensioner in nylon 46
(courtesy DSM Polymers).

Internal moving parts

Plastics components have for many years been associated with the smooth
running of metal parts. Examples are the use of glass reinforced PA 66 for
the oil deflectors on valve stems, and the use of PTFE for exhaust valve
stem seals, which need to survive at up to 180°C.

Plastic piston skirts have long been a development topic; the first succesful
example was the use (in diesel engines) of inserts in carbon fibre reinforced,
internally lubricated PES by Floquet Monopole in France. Two of these
in each skirt have the effect of reducing the piston-to-cylinder clearance,
with consequent reductions in noise, fuel consumption and pollution, and
improved performance (see Fig. 6.20).

6.20   Piston skirt in carbon fibre reinforced
PES, by Floquet Monopole (courtesy ICI).

Connecting rods have attracted a great deal of attention, because of the likely benefits of weight reduction within the moving mass. Con rods in carbon fibre reinforced composites were used in racing vehicles in the USA as early as 1978, and subsequent designs have been developed and evaluated by Volkswagen and by a consortium in the UK. These have certainly achieved quieter running, with reduced vibration and fuel consumption. There are doubts about the strength and fatigue resistance, however, and no doubts whatever about the excessive cost.

Composite engines

It has been demonstrated very clearly, starting with the Polimotor in the USA in 1980, that an effective internal combustion engine can be built from predominantly thermosetting plastic parts. Initially, the costs appeared to be totally unacceptable, but after much development work along the way in several countries, we are approaching the time when a 'plastic engine' could be commercial reality for some types of volume car.

The Polimotor in its current version (a joint venture between Polimotor Research and the Rogers Corporation) is made primarily of glass reinforced epoxy. Cylinder head, cylinder block, cam cover and sump are all plastic, totalling around 60% by weight of the whole. Many key components are metal, and likely to remain so: these include the exhaust valves, valve springs, the cylinder liners, the cylinder head threads and bolts, the cam shaft and cam bushings, and the crankshaft. Compared with an equivalent

6.21 BRITE composite engine installed in a
Ford Fiesta (courtesy DSM Resins).

cast iron unit weighing 160 kg and delivering 105 horsepower, the Polimotor
weighs only 79 kg and is said to deliver 175 horsepower at 5800 rpm.

Companies in France, Germany, Holland and the UK have collaborated
with Ford Motor Company to develop a composite engine in a BRITE
(Basic Research in Industrial Technology in Europe) programme. This is a
1 litre engine and can be seen in Fig. 6.21 fitted in a Ford Fiesta. The
construction basically uses a central core made from conventional metal,
with outer walls of reinforced thermosets bonded and bolted to it. The
choice of different composite systems for different parts of the engine is
interesting; it serves also as an indicator of the strength of the composites
'portfolio':

- Crank case and chain case were made in epoxy, reinforced with a continuous glass fibre random mat, which gives a largely isotropic structure.
- For the sump another epoxy was used, with easier processing but inferior

heat resistance, reinforced with layers of uni-directional glass fibre stitch-bonded together.

- The cam cover (visible in Fig. 6.21) was injection moulded in BMC (polyester). This is a potential weight and cost saving alternative to aluminium, when the numbers required justify the tooling costs.

On road testing the engine performed well, demonstrating quicker warm-up and 5% greater economy. Lower exhaust emission and much quieter running constituted two important social benefits. Oil leakages suggested that fundamental gasket redesign might be necessary. Weight saving compared with aluminium is potentially over 10% and compared with cast iron, over 30%. Many different composite inserts and resins could be used, but cost reductions (in raw materials and processes) will be necessary if any form of the 'plastic engine' is to attain volume production.

Suspension

The role of plastics in the chassis area is more important than is generally appreciated. Thermoplastic elastomers are used for MacPherson strut bellows, as anti-roll bar bushes, in front suspension coil spring assemblies, and as bellows and spring stops in shock absorber systems. Nylon and acetal are used in applications needing rigidity and low friction as well as resilience, such as guides and bushing in suspension arms and telescopic dampers. Acetal (homopolymer and copolymer) is often the first choice for unlubricated assemblies, because of the consistency of its frictional and wear performance over a wide range of temperature and moisture conditions. For the shock absorber spherical joint cap in Fig. 6.22, a high molecular weight grade was selected, for maximum impact strength.

Reinforced PTFE imparts low-friction non-squeak characteristics to seal-

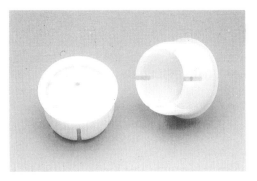

6.22   Shock absorber spherical joint cap in acetal (courtesy Du Pont).

ing rings in hydraulic systems and to the interleaving in rear spring suspensions. Small components of this kind are very widely used, and for the most part taken completely for granted. Material selection in this area can be difficult. It must start with an appreciation of the relative importance in the particular application of the key factors: creep, elasticity (recoverable strain), fatigue, lubricity and abrasion resistance.

A much higher profile development, which has attracted a great deal of research effort, is the concept of leaf springs made from continuous fibre composites. The attraction of using such materials for leaf springs is based on their anisotropy (the enormous longitudinal strength and modulus), and the resulting reduction in unsprung mass, with (in general) better roadholding. Interest in composite leaf springs has focused on small vans, where these benefits are particularly relevant, and where the production volumes of any one model variant tend to be small. Some very successful developments have been reported from companies such as GKN, but overall the extent of commercial exploitation has been disappointing. There are some technical problems: as with all composite components in a variable stress situation, the onset of fatigue is difficult to detect, and furthermore the fixture points usually need special strengthening with additional bushing, because the longitudinal advantages of anisotropy do not extend to multi-directional shear stresses. The lack of an accepted recycling regime has also acted as something of a brake on the development in recent years. None of these problems is decisive; the essential reason for the lack of acceptance of composite leaf springs may simply be that the cost incentives are insufficiently great at the present time.

In the long term, composites may have considerable benefits to offer in axle technology. TRW in Germany have developed a stabilizer linkage arm (effectively an anti-roll bar) for the Peugeot 605. This is an injection moulding in nylon 66 with 43% glass fibre. There are distinct improvements in respect of noise, damping, fatigue and vibration, and 10% cost saving and 20% weight saving are claimed. In more fundamental developments, both Audi and Daimler Benz have reported successful road tests on multi-link independent rear axles, including links made entirely from composites. Once again, doubts are expressed about the difficulties of identifying fatigue damage in its early stages, and of streamlining the production process to achieve acceptable costs.

Steering

The use of plastics in steering wheels is discussed in Chapter 4. In steering mechanisms plastics have a key role. Nylon is widely used for the housings of the lock mechanism at the top of the steering column, and for the stone

6.23   Boot for constant velocity joint on 1992
VW Golf, in thermoplastic elastomer (courtesy
Du Pont).

guard at the base of the column. PA and reinforced PA also feature in the various devices which have been designed to ensure controlled collapse of the steering column under impact. Acetal parts are used in most steering rack-and-pinion systems, and lubricated PES has been used for a spring-loaded plug which curbs fretting and rattling in steering racks.

Power steering assemblies include many plastic parts. Both glass reinforced nylon and acetal mouldings are used for reservoirs and for power steering fluid (which can reach temperatures of around 140°C), in containers made of two separate mouldings welded together. Unfilled PA 66 appears as pump caps and dipsticks, and reinforced PTFE as bushes, seals, bearings and ball-and-socket joints around power cylinders, racks and stub axles. Ball-and-socket joints in track rod ends are made from a variety of plastics pairings, notably acetal and thermoplastic PUR, and tie-rod ends are often coated with fluoropolymer. Figure 6.23 shows a boot for a constant velocity joint on the 1992 VW Golf, in thermoplastic elastomer. This material replaced polychloroprene synthetic rubber, on account of its superior low temperature fatigue resistance under Nordic winter conditions.

Brakes

The possible introduction of plastics brake pedals is discussed in Chapter 4. This possibility could be said to represent the ultimate intrusion of plastics into brakes technology. The 'intrusion' in fact started as early as the First World War, when PF resin was first used to impregnate asbestos brake linings.

The principal plastics application is in fluid reservoirs; these have been made from PA (reinforced and unreinforced) and HDPE, but nowadays are usually in injection moulded PP, heat stabilized. Improvements in blow moulding technology are reviving interest in blow moulded PA 6. Another very important use for nylon is in air brake lines, and hence essentially in heavy goods vehicles; these are extruded from PA 11 and (mostly) PA 12.

Glass reinforced PA is used to protect brake discs, in the form of stone and splash shields. More significantly, both PA and reinforced PBT have been used for the housings of brake servo mechanisms. Small quantities of more exotic plastics are used in sophisticated ABS systems; e.g., injection mouldable fluoropolymers for brake pad sensor housings, and PEEK for carbon brush holders.

Fuel tanks

The blow-moulded HDPE fuel tank is so firmly established in the motor industry today that it is hard to believe the prejudice against plastic tanks which persisted even up to the early 1970s. Partly this was the inertia effect of old legislation: when the first Volkswagen with an HDPE tank went into volume production in 1973, constructors in the UK, for instance, were still compelled by a law of 1929 to make tanks out of steel.

The change was stimulated by the realization that plastic tanks could be readily fabricated into awkward, space-saving shapes, and that, far from being more dangerous, plastic tanks were potentially much safer than metal tanks, because the risk of explosion was virtually eliminated. The choice of material was initially difficult. Of the plastics available at acceptable cost, nylon offered the best impermeability to petrol, but was extremely difficult to fabricate into large containers with adequate impact strength. HDPE grades of very high molecular weight quickly became the chosen material for the application, on account of its processability in large blow mouldings and its excellent impact strength (see Fig. 6.24). The impact tests are extremely severe. The European standard, ECE R34, includes a test with a sharp pendulum, a pressure test, and drop tests equivalent to 50 km/hr collisions.

Blow moulding is the ideal production method for all designs of fuel tank up to a capacity of around 100 litres, provided the numbers required justify

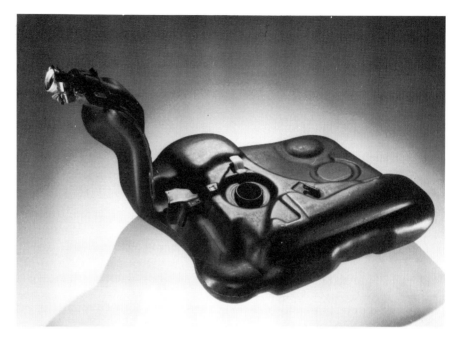

6.24   Typical fuel tank, blow moulded in
HDPE (courtesy BASF).

the tool cost. Rotational moulding however has an important role in the
market. Tooling costs are much lower, and there is much less material
wastage. Rotational moulding is used for very large tanks (up to 800 litres)
for special vehicles, and is popular for any application where the run length
is no greater than around 500 to 1000, and the shape is not too impracticable;
many heavy vehicles meet this description. Rotational moulding is also
extremely useful for making prototypes of designs which will eventually be
executed in injection moulding.

Additional processing stages have been needed to achieve the requisite
low permeability. Fluorination and sulphochorination are used, but
sulphonation is the most common method. This is sufficient to meet the
requirement of the European test, ECE R34, but not always enough for
the American SHED (Sealed Housing for Evaporative Determination) test
(FMVSS Part 86), which measures fuel escape from the whole car, and not
just the tank. Drastic steps sometimes need to be taken to pass the American
standard; Volkswagen for instance actually devised a system for locating the
fuel pump inside the tank. Rotationally moulded tanks for HGVs do not
normally need surface treatment, because diesel fuel is less volatile and less
likely to permeate, and the tanks are mounted outside the vehicle.

6.25   Fuel tank level gauge in acetal (courtesy
Du Pont).

Advances in extrusion and blow moulding technology are beginning to displace the unpopular surface treatment processes. Nissan introduced an expensive but effective five-layer process in 1987. This comprised a central layer of nylon 6, with adhesive on both surfaces, securing it to layers of HDPE. This HDPE-PA 6 combination proved to be satisfactory for methanol-based fuels as well as hydrocarbon-based petrol. Krupp-Kautex have demonstrated that as many as six layers can be incorporated in a single blow moulding operation, to ensure a particular range of properties, including impermeability. A nylon 6 layer, as before, can provide petrol impermeability, and a layer of ethylene-vinyl alcohol is an excellent barrier against methanol. Ford in the USA began small scale production of six-layer tanks in 1993. Du Pont have used a formulation route rather than a machinery route, by introducing flakes of nylon into HDPE. On blow moulding a surface layer rich in the impermeable nylon is produced. Fiat adopted this laminar barrier technology for all its 1993 models.

Petrol filler caps have been injection moulded in nylon for many years. The advent of methanol in fuels has caused most other fuel tank accessories, such as the filler neck and level gauges, to be made in acetal, in preference to nylon (Fig. 6.25).

7

Electrics

Ignition

Robert Bosch patented high tension ignition in 1902: this was the beginning of the long history of polymers in automotive electrics, culminating in the present situation where some 25% of the cost of a car is represented by electrical components. Rubber was the universally used cable insulating material until the arrival of PVC, and indeed rubber was for many years the only insulating material which could be shaped. PF resin was used in paper and fabric laminated insulation from 1909, but it was not until the mid-1920s that PF moulding technology had advanced enough for shaped mouldings to begin to take over from rubber, in components like magnetos and distributors. From then on developments in the PF moulding industry and automotive electrics were very much intertwined.

Polyester BMC was first used for distributor caps and housings and ignition coils in the 1960s, but thermoplastics did not begin to advance at the expense of thermosets until the 1970s. This process has extended less in Europe than in the USA, where PET and PBT are now in use for distributor rotors and caps, coil housings and heads and spark plug caps. Nylon is represented as mineral filled PA 66 for coil bodies, and more recently PA 46 for spark plug covers. Grease-free moving parts for Mercedes ignition switches have been reported in a special PA 66 composition incorporating PTFE and aramid fibre. Magneti Marelli have introduced special materials for high performance end uses, such as PEEK in distributor rotor arms in Ferrari Formula One cars (see Fig. 7.1).

Battery boxes

Battery boxes were made from very heavily filled thermosets until relatively recently. Apart from the use of polystyrene for battery boxes in the NSU Prinz in the early 1960s, for a battery installed in the passenger compartment, thermoplastic battery boxes did not achieve general acceptance

148

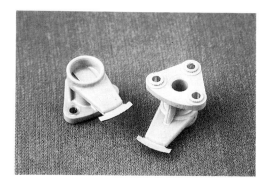

7.1   Distributor rotor arm in PEEK for high
performance car, by Magneti Marelli (courtesy
ICI).

until the early 1970s. Since then, polypropylene has taken over almost
completely, being by general agreement the ideal material for the application, by virtue of acid resistance, mouldability and low cost. Early
problems with dimensional control have been resolved by the advent of
'controlled rheology' grades with virtually isotropic shrinkage behaviour.

Circuitry

Connectors

Something like 50 billion connectors are manufactured in the world each
year. They are essential to every kind of electrical component and appliance,
and modern technological society would be inconceivable without them.
Nearly all of them involve a thermoplastic injection moulding. There are
on average about 200 in every car, which means that the motor industry
accounts for one fifth of the world connector market.

The 'workhorse' for connector moulding is unfilled nylon 66, because of
its fitness for the underbonnet environment, and its useful combination of
toughness and creep resistance, making it ideal for snap-fit assemblies which
maintain positive contact. The formulators have had to respond to tightening
specifications, particularly in respect of flame retardancy; as a result there
are grades available now which achieve the UL 'V-0' rating without losing
the essential 'snap-fit-ability'.

Increasing sophistication under the bonnet usually means new opportunities for plastic connectors, with ever harsher temperature conditions,
beyond the capability of PA 66. This is an area where several performance
materials compete, notably PPS, PES, PEI, PA 46 and semi-aromatic PA.

7.2   High temperature connector in PES
(courtesy ICI).

Selection has to be based on some fairly intricate differences in balance of properties, as the following examples show. A connector made by AMP for the BMW engine management system uses unfilled PES (Fig. 7.2); the reasons are the temperature peaks of up to 180°C, plus the need for resilience (associated with unreinforced materials), and uniform electrical performance over a wide temperature range (a characteristic of amorphous aromatic polymers). A connector by TRW for a VW automatic transmission system needs not only the ability to survive occasional exposure to 180°C, but also resistance to hot brake fluid; in this case a reinforced PA 46 was the best answer. Nylon 46 is favoured in other cases where the connector is in direct contact with the cylinder block.

Other moulded components

Most switches are necessarily located in the passenger compartment, and although they are not exposed to extreme thermal and chemical conditions, they do need to be robust and resilient. Nylon and acetal have long been accepted for most instrument panel switches, and the more complex multi-point switches located on the steering column. The same materials are generally used in micro-switches. Nowadays, switches in high temperature underbonnet locations are often in PA 46 or PPS. Plunger-operated switches for brake lights and door and boot lights, operating in dirty, wet or greasy conditions, are commonly made in acetal (see Fig. 7.3).

Underbonnet junction boxes are very visible and vulnerable components. A typical VW 'Zentralelektrik' is moulded from PA 66 reinforced with very short glass fibres (to give a combination of heat distortion resistance and isotropy), with a transparent cover in PC. Other designs use PP for the basic housing, with the more expensive nylon confined to the socket frame. PBT

7.3 Plunger-operated switches in acetal homopolymer (courtesy Du Pont).

is very widely used for such engine compartment electrical housings in the USA, and increasingly so in Europe.

Cables

The automotive industry was dependent on rubber cable insulation throughout its formative years. Cable sheathing with extruded PVC began during the Second World War, focusing on the needs of military aircraft; it was therefore some years before the motor industry achieved its present state of dependency on PVC.

The need to organize the layout of underbonnet cables, and indeed to protect the low-softening PVC from the increasingly severe temperature hazards, has stimulated other plastics applications. Cable ties in PA 66 (chosen for its combination of toughness and creep resistance) are injection moulded in their millions, to a wide variety of size and strength specifications

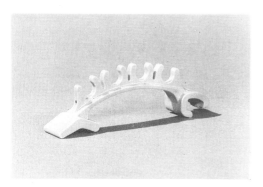

7.4   Cable guide in PES for Saab (courtesy ICI).

7.5   Rear light lens in PMMA for Ford Escort (courtesy ICI).

(see Chapter 2). Increasingly crowded engine compartments and the growing number of heat sources have led to the use of specially designed cable channels and cable guides, in more temperature tolerant materials such as PES and PA 46 (see Fig. 7.4).

The trend in the future will be towards miniaturization of many of the cables and connectors now in use. Multiplexing is the likely first stage, involving much less cable, but with higher performance insulation capable of surviving in much hotter locations. Fibre optics will follow when the cost of manufacturing and installation are suitable.

Lighting and instrumentation

Rear lights are complex assemblies of mainly plastic parts. Almost everywhere, the lenses themselves are injection moulded from polymethyl methacrylate (PMMA, see Fig. 7.5). The optical requirements for the separate functions: tail lamp, stop light, indicator, etc., are all different. The designs for the inner surface contours are intricate and the specifications are demanding. Design and production of rear light lenses has become a specialized industry, with the advent of two-colour, three-colour, and now four-colour injection moulding machines; the last to meet the current fashion for smoke-grey anonymity. The instrument backs are made either in ABS, which can be directly hot-plate welded to the acrylic lens, or in PP, which is cheaper but has to be assembled with screws. The lamp sockets are usually in glass reinforced PA, PET or PBT.

Several plastic materials have been used for headlamp bodies. Glass reinforced PP and BMC are most widely used, although one of the largest, in the Opel Senator, uses SAN. Some current designs combine the housing and the reflector in a single moulding; most materials can now be moulded to give a surface sufficiently smooth to be silvered without needing a pretreatment. The very low shrinkage of unsaturated polyester means that, in the form of injection moulded BMC, it is the most popular material for precisely curved reflectors.

Once again, temperature requirements are rising, and with them the materials specifications. The causes are the advent of halogen lamps and stylistic influences leading to smaller lamp bodies. Systems such as PES reflectors inside glass reinforced PA bodies have appeared, and PEI and PPS are both used in different reflectors for BMW. Volkswagen are using reinforced PPS for fog lamp relectors, and the same material has been envisaged for a single reflector plus housing moulding. Many headlamps feature a moulded surround in glass reinforced PA, and to pursue the ideal of a completely corrosion-free lamp, adjusting screws have been designed in long fibre reinforced nylon.

Plastic headlamp lenses have only recently become legal in Europe, and their use is now incorporated in ECE R20. However, plastic headlamp lenses have been legal in Japan for many years, and in the USA since 1979. The appropriate standard there is FMVSS 108, whose heat distortion and scratch requirements are met only by polycarbonate, with a silicone hard coating. A plastic lens only weighs half as much as glass; the main incentives however are the lower assembly costs represented by plastics and the much greater design freedom, especially for 'flat' low aspect ratio lenses, suitable for low-front aerodynamic bodies.

The use of plastics in interior instruments is covered in Chapter 4; it is worth noting that major changes in the style and content of instrument panel display are likely to accompany new electronic functions and display styles. These will still need a housing and a transparent protective cover, however, so the present materials are unlikely to be displaced in the short term.

Other electrical equipment

A very large catalogue of 'miscellaneous' applications for engineering plastics could be assembled, extending into every piece of electrical equipment. A few generalizations are all that can be given here:

- For gears in such items as power windows and windscreen wipers, acetal is usually the first choice. Unfilled nylon is also very much in evidence; in gear trains in fact there are positive advantages in using the two materials together for alternate gears.
- Housings for alternators, small motors and items such as rear wiper gear boxes are usually made in glass fibre reinforced engineering plastics: simplistically, PP for low cost, PA 66 for higher temperature and tough-

ness, and PET for long-term heat resistance and dimensional stability in varying moisture conditions.

- Internal motor components such as half rings, brush holders, insulators and bobbins are very frequently found in reinforced nylon 66. Reinforced PET may again be preferred for long term stability, and for higher temperatures the choice may be nylon 46 or PPS, PSU or PES.
- Several materials have been installed as the working parts of windscreen wipers, but the most successful in terms of consistent performance in all climates for exposed parts like the wiper holders and the wiper supports appear to be reinforced PET and PBT.

Electronics

Electronics are expected to contribute as much as 20% of total vehicle costs in the private car sector by the year 2000. This is an increasingly important application area for plastics: not so much in terms of volume (in fact the 'electronics revolution' is often characterized by miniaturization), but in terms of the opportunities it provides for high performance materials across the spectrum from traditional engineering plastics to 'speciality' polymers.

This chapter includes descriptions of application areas such as connectors and switches. These items, being small components produced in large volumes, need injection moulding materials with fast cycling ability. This has long been a fertile growth area for PA 66, and more recently for PET and PBT. Higher ambient temperatures in the engine compartment are turning attention increasingly to materials with continuous use ratings in the range from 150 to 200°C, such as PSU (polysulphone), PES, PAA, PPS, and PA 46. Amongst these high performance materials, the amorphous polymers like PES and PEI in the unfilled form are best at combining 'snap fit' and resilience with good heat deformation resistance and tight dimensional tolerances, whereas highly crystalline polymers like PPS and PA 46 ensure the best creep and deformation resistance and high temperature chemical resistance.

Exactly the same kind of factors determine the material choice for the new electronic control systems. The role of the plastics material is to provide encapsulation, or a housing, or a circuit board component, keeping the conductors separate and functioning over a wide temperature range, often in the presence of hostile chemicals. A sensor, for instance, may be located permanently within a rear axle, like the AB Elektronik speed sensor for BMW. The part of the encapsulation which is in permanent contact with the hot oil, at up to 160°C, is in PA 46, while the less vulnerable part is moulded in PBT (see Fig. 7.6). Other sensors, with temperature exposure confined to the −40 to +130°C range, used for so called 'drive-by-wire' systems con-

7.6 Sensor encapsulation in nylon 46 (courtesy
DSM Polymers).

trolling fuel injection or active suspension for instance, will most commonly use reinforced PA 66. Again, the higher temperature needs are often being met by PA 46 and PPS.

Surface mount technology is being used increasingly in assemblies such as in-car telephones and audio equipment. PBT and PET, more stable dimensionally than PA 66, are widely used. Distributorless ignition systems (DIS), for example, use a tray in PBT, with surface mounted conductors, filled with epoxy and encapsulated with PET. Vapour phase soldering can involve temperatures of up to 260°C for several seconds. This is encouraging the use of fully aromatic polymers like PPS and PEI, which not only have a high HDT but are also inherently non-flammable.

There is a small but growing specialist market for conductive plastics. In terms of surface resistivity, these offer a middle ground between the two extremes of conducting metals and insulating plastics. There are two main categories: the first comprises anti-static compounds which are sufficiently conducting to dissipate surface charge, and so avoid the safety and damage hazards of electrostatic discharge (ESD). More recently, grades of much higher conductivity have been designed, as EMI attenuating compounds to protect electronic equipment from electromagnetic and radio frequency interference. These compositions are usually based on polymers such as ABS, PVC, PC and mPPO, most commonly reinforced with fine fibres or powder or flake, in stainless steel, copper, aluminium or carbon. Housings made from these materials compete with mouldings in conventional grades which are subsequently painted or flame sprayed with a conducting layer.

8

Recycling

Recycling: an unavoidable issue

Recycling is now a major factor influencing the use of plastics in the automotive industry. This was not always the case. Until the surge in environmental awareness began in the early 1970s, the plastics industry (in spite of its excellent record in handling in-line scrap) displayed no interest in recycling its post-consumer waste products. Meanwhile the motor manufacturers confined their interest to absorbing a proportion of recovered steel from an inefficient and 'down market' scrap metal industry.

The instrument for change was the reclamation industry itself. The first stage was the development of high technology plant for the rapid shredding of cars: the modern shredder (the first in the UK was installed in 1967) can process 500 cars in an hour to yield 350 tonnes of ferrous metal. (Thirty years ago this operation would have taken a week.) The second stage was the gradual realization that, because of the growth of plastics in vehicles, this high-investment, high-technology shredding industry would soon become unprofitable. This burgeoning plastics content has a double negative effect: not only is the yield of valuable steel scrap declining, but the yield of unsaleable shredder waste is rising, at a time when the only effective disposal route, landfill, is becoming much more expensive and much less acceptable.

The alarm signals came first from Germany, where the steel industry (already ill-disposed towards plastics) had the most to lose. The problem is viewed with different degrees of urgency in different countries: it is at its most extreme in Germany and Holland, where the objections to landfill are the greatest. In other countries the priorities in environmental problems are different: France, for instance, is more concerned with energy conservation, and Italy with atmospheric pollution. Less affluent countries are more preoccupied with economic survival than environmental problems. Since the late 1980s however, the 'bottom line' message coming out of the EC in Brussels is that something must be done urgently about recycling automotive plastics.

Proposals that using plastics in cars should be curbed by legislation have never received much support, because the automotive industry is too well aware of their advantages. However, with the plastics content having risen from 2% by weight to something like 12% in 30 years, and rising still, the situation is urgent. The largest shredder in the UK, the Bird Group, forecast in 1990 that their margins would move from profit into loss in 1995.

Thus it was that by 1993 the combination of industrial and environmental pressures had made the recycling of plastics into a matter of major concern. The importance of this issue is undeniable; nevertheless it is also true that there are several ultimately more serious problems of resource management which have received far less attention.

The scrap problem

Prior to the shredding stage, procedures for collecting, stripping and compressing cars are variable and somewhat uncontrolled. Identifying and removing reusable components only happens in places where there is an organized infrastructure. Very often all that happens before shredding is draining the fuel tank and crushing and partly shearing the vehicle to make it easier to transport in bulk (Fig. 8.1).

Shredding reduces the vehicle to fist-size fragments in a matter of minutes.

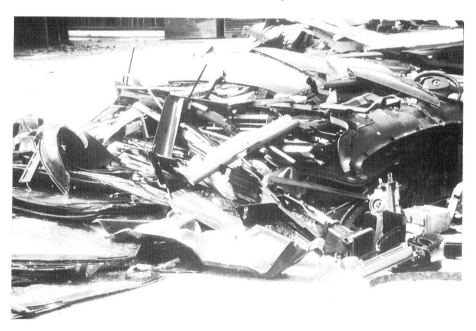

8.1   Sheared car fragments, prior to shredding (courtesy Bird Group).

8.2   Shredder waste (courtesy Bird Group).

The prime product is the steel, which is magnetically separated and then cleaned. The mixed non-ferrous metals are cleaned and resold. The residue is 'fluff', in light or heavy fractions, made up of glass, fabrics, rubber, lubricants and miscellaneous dirt as well as plastics. Conventionally, both categories of fluff have ended up as landfill. The systems as designed therefore make no provision whatever for the recovery of polymeric material. Meanwhile the mountains of fluff accumulate (see Fig. 8.2).

Many organizations are attempting to address this problem. All are agreed that some kind of dismantling operation needs to be put in place before the shredding stage. Ideally, there should be a system which recovers usable components intact, and which recognizes individual polymers and keeps them segregated. However, superimposing these requirements onto a world-wide scrap rate of over half a million cars per week, in shredding operations which yield over 100 tonnes of product per hour, presents a logistical problem of huge dimensions.

Pilot schemes for dismantling and sorting operations which focus on plastics have been in operation since 1989, and much progress has been

made. The underlying equation of the scrap industry is unchanged, however –
the revenue gained from the recovered material must exceed the costs of
recovery. Numerous routes have been suggested towards balancing this
seemingly hopelessly unbalanced equation.

Revenue can be gained by converting plastics to energy; most after all
have a much greater heat content than coal. However, investment is necessary for this to be done efficiently and acceptably. The main source of
new revenue must be the recovery of reusable components and reusable
materials. However, the traditional methods of recovering components and
identifying polymers (let alone segregating them!) are far too slow to be
compatible with modern shredding operations.

The other half of the equation i.e., reducing recovery costs, calls for
different solutions. In some countries, such as Germany and Sweden the
cost of collecting a scrap car is in effect transferred to the car owner, who
can only recover its residual value (and be relieved of the obligation to pay
tax) when the car has been conveyed to the scrap dealer. The big question
remains: how to set up a high speed, effective plastics stripping and segregating operation, compatible with sophisticated shredding, without impos-
sibly high costs.

The following section looks at the extent to which the cost effective
recovery of polymers and plastic components might be achieved.

The alternatives for plastics

There are several quite separate issues involved in plastics recovery; all must
be adequately dealt with before a solution can be achieved.

Recovery of components

The refurbishing and resale of worn parts is of course an old tradition of the
automotive industry. In some countries there is a well developed infrastructure extending this custom to plastic parts. It is likely to be confined to
damaged cars, or vehicles taken out of service prematurely. There is not
expected to be a big market for plastic components from cars scrapped after
10–15 years service. The main motivation in dismantling plastic parts intact
is to recover sources of unmixed polymer, in a relatively clean and pure
condition.

Recovery of 'pure' polymers

Naturally enough, vehicles have been designed for some permanence, with
little thought for rapid dismantling. This however, with the implied ability to
separate connected parts cleanly, is the key issue. One of the first studies

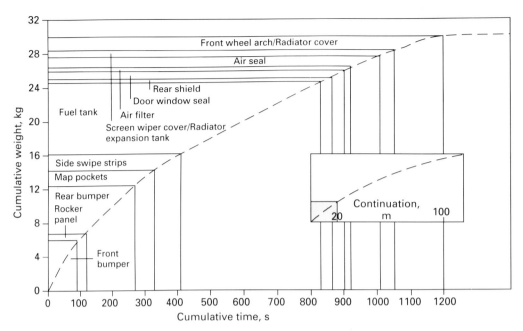

8.3   Diminishing return effect in recovering polymers from scrap cars (derived from work of Forschungvereinigung Automobiltechnik; data supplied by BASF).

of dismantling was carried out by Porsche for the German Automotive Engineering Research Association, on the 1983 VW Passat. It was found that 63% of the total 97 kg of plastics could be sorted, with 30% of the total weight being dismantled within 20 minutes. Beyond the bumpers, fuel tank and larger trim pieces, there was a clear diminishing return effect. A further 54 minutes was needed to remove the remaining components, with another 41 man minutes needed to dismantle these components themselves. This totalled 71 seconds per kg of plastics in which to dismantle the components deemed to be 'sortable' and separate them into classes of polymer (see Fig. 8.3).

More recently, work by Ford on dismantling the Escort confirmed the diminishing return effect; 20 kg of recyclable plastics were recovered in the first 10 minutes, but only 10 kg in the next 10 minutes. Dismantling costs were calculated as about 50 p/kg after 20 minutes, rising to about £12/kg after 50 minutes.

The challenge for designers is to design vehicles which can be readily dismantled, without compromising the safety or integrity of the vehicle in service. This could mean discarding some of the advantages of modern adhesives and welding techniques, but this is a questionable course to take. Whatever changes are made, it seems unlikely that more than about 30% of the plastic content in a vehicle can be recovered cost effectively in an uncontaminated state.

The most satisfactory way of recycling polymer recovered from 'clean' components is to regranulate the material and use it again in the same application. In this way the effects of paint or surface deterioration are minimized: this has been demonstrated in many cases, e.g., Volkswagen bumpers in polypropylene.

Polyolefins in fact lend themselves to this straightforward recycling procedure. Other thermoplastics like polyamides and polyesters need more careful handling because of their sensitivity to moisture absorption and molecular degradation. Thermosets are, by definition, not recyclable in the true sense; however by comminuting and recompounding with fresh monomers, most thermoset polymers can in effect be recycled without significant loss of performance.

The conclusion is that nearly all polymers recovered from identified 'pure' sources can be recycled, in admixture with virgin polymer, albeit with the likelihood of a small deterioration in performance. It has been calculated that the average car contains 17 different types of polymer. However, it does not end there, as there are likely to be as many as 60 different grades represented. At this point the logistics can become very complicated. Rather than attempt to segregate each and every variety of PP or ABS, etc., it makes more sense to segregate only according to generic type, and not according to precise grade. The end products can be grades of mixed recyclate, with a single polymer base, and while their property balance will be different from standard grades, they are still capable of being subjected to product specifications.

The 'cascade' concept has gained much credibility; the term covers the successive recycling of a plastic material, accepting that there will be progressive slight deterioration, without the loss of the essential properties of the material. At each stage of the cascade the specification is reduced and the application focus becomes less critical. However, the cascade idea only makes sense when applied to engineering plastics which can command a relatively high price. Figure 8.4 shows a good example; this is a roof spoiler carrier, moulded from recycled bumpers in PC/PBT alloy.

There is no 'right' way of proceeding which is universally applicable or acceptable. Sensitivity to contamination varies with property and with polymer. Furthermore, what is acceptable depends very much on price. The higher the price bracket of the original material, the greater the incentive to recover and recycle. For bulk polymers, however, the economics of recycling are fraught with difficulty. If the price of recyclate is too high, there is no

8.4   Opel roof spoiler, moulded from recycled
PC/PBT bumpers (courtesy GE Plastics).

incentive to use it in place of virgin polymer; if the price is too low, there is
a risk of disrupting the market for virgin polymer (on which the recyclate
ultimately depends!).

Regeneration of monomers from recovered components

The regeneration of monomers from recovered components has been included for completeness, but in reality the concept of 'chemical dismantling'
is unlikely to be realized on any large scale. There are several possible
options for the chemical recycling of polyurethanes. Polyesters and polyamides can be broken down into their chemical parts by hydrolysis, and
pyrolysis can be used to recover PMMA and polyolefines. There are problems of scale, however, and possible atmospheric pollution problems. This
is not a solution for the present time, but the technology is potentially
available, against a time when petroleum becomes much less available and
much more expensive.

Recovery of mixed plastics

Much has been written about the endless application possibilities for mixed plastic waste. Fence posts, floor tiles, storage pallets, road cones and other large volume, low technology items are cited, and there have indeed been some successes. The fact is, however, that randomly mixed polymers can be difficult (and possibly dangerous) to process, with a variability in perform-ance which could be unacceptable even in these undemanding applications. Furthermore, the size of this 'market sink' is not unlimited, and it is arguably better served by second grade but unmixed bulk polymers, especially in various reworked and somewhat contaminated manifestations, at very low prices.

In summary, there is clearly a market for clean, pure recyclate, and a market for recyclate which is essentially single polymer, but with known contaminants in not-too-large quantities. As a saleable, manageable product for plastics applications, the random polymer mix looks a poor proposition. Its value as a plastic raw material is almost certainly less than the value of the energy it contains. And therein lies the solution.

The energy aspect

Research into methods of recovery and segregation of polymers is proceed-ing vigorously, but it is too early for definitive conclusions. Nevertheless it looks very unlikely that it will ever be cost effective to separate (into materials of defined specification) much more than about one third of the plastics content of a car. Laws enforcing limits on the range of base polymers employed (a thoroughly retrograde and restrictive strait-jacket in itself) will not make much difference, because of inevitable grade variations and the inclusion of metal parts in sub-assemblies, and the remorseless logic of diminishing return economics. There will still be four or five million tonnes of shredder waste arising in Europe each year, having no value as a plastic raw material, and facing increasing objections to its disposal as landfill.

Polymers, however, have a very high heat content. Even 'fluff', hopelessly mixed and contaminated, has a higher calorific value than coal. Energy recovery is therefore a very proper solution, for this residual non-metallic material. New technology is needed, to handle huge volumes of low density solid fuel; this is in development, with the shredding and recovery industry taking the lead. An 'integrated energy' plant to generate electricity from shredder waste has been designed by the Bird Group, in collaboration with SGP-VA of Austria and Best Global. The new plant is designed to meet very rigorous emission standards, with sophisticated fail-safe procedures.

Reliable technology for the safe, high temperature incineration of polymer waste is now proven, but the task of overcoming the legacy of ignorance and misinformation about the subject has barely begun.

Recycling composites

Obsessive focusing on recycling can distort the whole picture of automotive design and construction. The logical end to simplifying material choice would be to make all components from one material; and that of course would not be a composite. But it is now abundantly clear that new design and material benefits in the industry, particularly the enormous savings in energy overall, are largely dependent on plastics, and composites especially. Compromises, therefore, are inevitable.

All material admixture complicates the recovery process. Paint on car bodies creates problems in the foundries; but no one is suggesting that steel should not be painted. Hence the issue of composites recovery cannot be avoided. In fact the situation is not fundamentally different from that with 'pure' plastics; it is just rather more complicated.

When unfilled polymers are repeatedly recycled, there are in general two factors causing the properties to deteriorate. Firstly, the molecular weight is reduced because of the additional thermal history at each recycling, and secondly each handling stage adds some contamination. Toughness, colour, and general reliability are progressively affected. Composites however are less sensitive to this deterioration, because the reinforcing fibres dominate the mechanical behaviour, and (to a degree) will literally 'hold together' a compound in which the polymer has become quite fragile. However, those properties which are determined by fibre length (which are probably the ones providing the performance 'edge'), will suffer at each recycling.

Composite scrap still represents polymer with enhanced performance, and on the other hand only contains around 50%–80% of the recoverable energy of the polymer. Hence in the marginal zone of the recovery process with its 'diminishing return' curve, composites may make more sense as recovered material than as recovered energy. Furthermore, long fibre scrap is more useful than short, in any cascade system of recycling.

Composite recyclate may no longer be suitable for its original process. SMC scrap can readily be reused, suitably ground at around 20% of the whole, in new sheet moulding compound. This is now happening with the longest running SMC car panel application, the Chevrolet Corvette. This kind of addition, of randomly oriented short fibres in a controlled proportion, can work well with BMC, and (with rather more difficulty) with the various types of glass mat reinforced thermoplastics. Scrap GMT-PP, being thermoplastic, can in theory be 100% recycled into injection moulding compound. Such material has similar properties to glass reinforced PP made by extrusion compounding virgin material, but because of fibre length variation may be less consistent.

In summary, composite scrap of any type is likely to have interesting properties, which however may not be as predictable as we would like, especially if fibre orientation is important to the component function. Mixed composite scrap could be extremely useful in applications where stiffness rather than strength is being sought. The sandwich process in injection moulding could be particularly suitable, with the unseen core providing great rigidity at low cost.

The pressure on designers

Overview

Recycling was, at the time of writing, the dominant issue in the automotive industry, but as has been shown, this situation arises from the impending crisis concerning economic viability in the shredding industry. In the background are two potentially more serious issues. The first is the increasing scarcity and cost of fossil fuels (and let it not be forgotten that only 4% of 'the oil barrel' is used to make plastics, while some 85% is expended in providing energy). Secondly, there is the environmental damage due to exhaust emission, especially (because there is so much of it) carbon dioxide.

The average life of a car is 10 to 12 years. However, this period is stretched at both ends; adding on the design and development period at the beginning and the extended 'tail' from some car owners gives a total span of at least 25 years. This means firstly that the mistakes of a designer can survive for all this time, and secondly that the emergency recycling measures now being contemplated may be less relevant when today's designs come to be scrapped. Both of the big background issues demand urgent reduction in fuel consumption; these arguments can only get stronger. The message for the designer is therefore to press ahead with the appropriate application of plastics, with all its energy efficient benefits. He (or she) should make sensible provision for easier dismantling of vehicles, but not allow the recycling pressures to distort his overall balanced, holistic view of automotive design.

Joining components

Easy dismantling favours an assembly which is a push-fit, without adhesive or weld. However this must be balanced against the obvious economic advantages of modern adhesives and welding systems and their importance in robotic assembly, and the probable need to enforce closer tolerances in order to make push-fits reproducible and dependable (see Chapter 2). The main argument is against joining dissimilar materials; certainly the designer must bear this in mind, but he may be in danger of ignoring an even older axiom, telling him to choose the best material for the job!

The ubiquitous nylon fastener probably has a part to play here. Existing design capability already allows a choice between permanent locking and repeated disassembly: even more ingenious designs may be needed to ensure that assemblies are totally reliable until the moment when they are instructed to fall apart!

Identifying materials

Much has been made of the impossibility of differentiating one piece of black plastic from another. Coding systems have now been devised to remedy this: it remains to be seen whether these are compatible with each other, and still adequate in 15 to 20 years. Furthermore, how detailed should the identification be? Marking systems really belong to the first, labour-intensive stages of the stripping process, when large, well-recognized components are being detached. For later stages with higher speed stripping the markings would need to be machine-readable. But then questions arise about what constitutes 'a component'. The conclusion is that this is not a good road to go down, thus endorsing the view that, after the larger parts have been removed, the remaining two-thirds or so of the polymeric content is best left unsegregated.

Single polymer sub-assemblies

As designers have come to appreciate the versatility of plastics, and as pressure grows to simplify assembly line operations, many off-line sub-assemblies have become more sophisticated. Soft front ends and rear ends, and also high series instrument panels, are made up of three essential parts: the structural base (strong, rigid, dimensionally stable), the absorbing medium (flexible, highly deformable, energy absorbing), and the skin (impact and abrasion resistant, slightly deformable, cosmetic). There is strong pressure to design these sub-assemblies using materials with appropriately individual properties but all based on a single polymer.

This argument has considerable strength, because these are the components on which any meaningful segregation process must be based. Successful solutions have been designed in PP and PUR, but other polymers could also be used, at rather greater cost. There needs to a viable glass mat or long fibre reinforced material available (such as GMT-PP or S-RIM PUR), and an effective foaming capability, as well as an elastomeric or elastomer-modified skin grade with the right surface characterisics. Polyurethane is arguably in the strongest position from this viewpoint, because there are well-developed PUR adhesives and paints available, as well as foam and glass mat reinforced material.

New recycling initiatives

The recovery and recycling of plastics scrap is the subject of intense investigation in many countries. Only about 7% of the world's plastic scrap is automotive, but as will have been noted, the acute economic crisis in the car shredding industry ensures that the automotive sector receives a disproportionate amount of attention. Europe is taking a lead, partly because the environment is a high profile political matter, but also because the availability of landfill sites is reducing very rapidly, for social and economic reasons as well as geographical ones.

In the UK, the focus of attention has been the joint venture set up in 1991 between the Bird Group and the Rover Group. This aims to devise systems for environmentally 'friendly' recycling of cars, and to design for recycling. The picture is changing very rapidly, so it would be inadvisable to attempt an up-to-date survey here. The main focuses of attention are these: (1) Systems for stripping and segregating plastic parts, (2) Centres for doing this, and (3) Plants for the acceptable incineration of shredder fluff, coupled with energy recovery. Raw material suppliers, motor manufacturers and the shredder industry are all heavily involved.

9

The future and automotive plastics

Design trends and legislation

Recycling is currently the biggest issue in the automotive industry, but the other big issues, with potentially even greater environmental 'time-bomb' implications, are fuel consumption and exhaust emission. Legislative pressure is already being exerted on the latter two in several countries, and there can be no doubt that it will increase everywhere, on all three issues. Laws concerning excessive noise (a lesser concern, but a more 'visible' one!) are already in place, although not always enforced. Ideally, the enabling technology precedes the legislation, but this is not always the case (*vide* the American Corporate Average Fuel Economy (CAFE) rules which imposed progressively severe fuel economy targets on automobile manufacturers in 1978). There is a real danger in Europe that a proper concern for the environment, belated but widespread, will result in both the automotive industry and the plastics industry being seriously embarrassed by ill-conceived legislation. At many points today the plastics industry is presented with both challenges and opportunities.

Supplier involvement

Supplier-customer relationships in the automotive industry experienced far-reaching changes in the 1980s; the principal root causes were both Japanese in origin. The first was the so-called 'Quality Revolution' (originally conceived in the USA but nurtured in Japan). This aimed at nothing short of fundamental changes of attitude, signalled by such concepts as 'Right First Time', 'Total Quality Management' and 'Zero Defects'. All kinds of products and services were affected, but plastics came under particular scrutiny because of the proliferation in plastics components (somewhat in advance of in-depth knowledge of polymers within the motor industry). The second was the growth of 'Just-In-Time' (JIT) supply systems, developed from procedures of inventory control and waste elimination associated with the Japanese word 'Kanban'. These changes have resulted in very much closer

168

relationships between suppliers and customers, with responsibility and involvement extending right back up the supply chain. There are fewer suppliers, but all are highly committed.

For plastics applications this cultural change has meant in some cases that component production has gone 'in-house', in the Japanese or American mode, and in other cases a greater specialization on the part of suppliers. It has also induced greater reliability in the performance of plastic components, with a consequently greater confidence about plastics among automotive engineers. An important feature of the new culture has been the growth in engineering specifications for polymers and for plastics components. These can be expected to assume even greater importance as the pressure builds up to recycle automotive plastics, with all the attendant risks of defects.

Safety

Safety legislation has of course provided challenges and opportunities for plastics for a very long time. (And very often the availability of plastics has stimulated the legislators into stepping up the legislation!) Plastics have played a big part in enabling the requirements of the European and US Federal standards to be met, for interior and exterior impact. The great advantage of plastics is that the requisite mix of deformation resistance and energy absorption can be brought together in the same assembly.

The transformation of the car interior over the last 20 years has been closely involved with safety legislation. The introduction of resilient plastic skin materials with foam backing, together with sophisticated nylon and acetal-based seat passenger restraint systems, has delivered all that has been asked for, at the present level of 'passenger packaging' requirements.

The new event in the passenger compartment is the arrival in big-series models of the air bag, using several engineering plastics. In 1993 the air bag was fitted as standard across the BMW, Mercedes and Jaguar ranges, and in the Ford Mondeo. The practice is certain to extend to smaller cars, where in fact it should have a proportionally greater effect in injury prevention.

The trend towards ever lighter cars has been reversed in recent years, because of the current strengthening of the passenger cage, particularly with the wider use of intrusion bars for added lateral protection. With the need for fuel economy more urgent than ever, this is likely to encourage further weight-saving plastics applications.

Fuel economy and the energy equation

A comprehensive study of the energy equation shows that, far from being merely the 'least worst' option, plastics are in fact the best option. This conclusion results from an investigation of 'Ecobalances', pioneered by Dow Plastics. In simple terms, this looks at a material, or an assembly such as a motor car, in a holistic or 'cradle-to-grave' fashion. Comparisons are made on the basis of total life cycle energy, i.e. a summation of the energy required to produce the raw material and fabricate the assembly, plus the energy consumed in service, plus the energy expended in recovering the scrap, less the convertible energy retained in the scrap. It can be revealing to apply this method to the scrap recovery process itself; all too often it is found that the total energy used in recycling exceeds the energy content of the recyclate.

Polymers score well against metals in this comparison, because less energy is consumed in their production. However, for automobiles, the energy expended in operation is far greater than the energy expended in production. Hence, the saving of even a small proportion of the operating energy is significant in terms of the complete equation. The arguments for weight saving are very strong indeed, although they do not seem to have been a primary motivator for motor manufacturers until recently. In fact, a 3% weight saving approximates to a 1% fuel saving, so that replacing metals by 100 kg of plastics provides a fuel saving of at least 5%. Measured on the urban cycle the saving is much greater, as frequent acceleration causes higher fuel consumption.

It can be demonstrated that the energy saved in service, by using plastics to replace a heavier part, is often greater than the energy expended in making the new part in the first place. Table 9.1 introduces the concept of the 'pay-back', in considering the example of a fender (wing) made respectively in steel, aluminium and R-RIM PUR. Comparisons are made on the basis of a representative car which burns 10 litres per 100 kilometres, over a life of 150 000 km.

Column 2 shows the weight saving compared with steel. Column 3 translates this saving into fuel saved over 150 000 km, on the basis of 9 litres per kg, or an energy equivalent of 290 megajoules (column 4). Column 5 gives the energy invested in producing the respective parts. Column 6 indicates that only in the case of the plastic fender does the energy saved exceed the energy invested, and that on a life of 150 000 km, the pay-back is realized within 40 000 km.

Many factors combine to improve the standing of plastics in energy terms. Considerations like reduced wastage in manufacture, simplified assembly through component integration and lower drag coefficients all play a part. Aerodynamics are important, at least in vehicles habitually driven at high speed. The Opel Calibra has a quoted drag coefficient of 0.26; Opel have calculated that without plastics, the best possible figure would have been 0.31.

Table 9.1. Fender materials: energy savings

	Fender weight, kg	Weight saving, kg	Fuel saving, litres	Energy saving, MJ	Energy invested, MJ	Pay-back distance, km
Steel	5.3	–	–	–	285	Never
Aluminium	2.9	2.4	21.6	700	850	Never
PUR R-RIM	2.5	2.8	25.2	812	220	40000

(Data supplied by Dow Plastics)

There are many ways of making the calculations, and many ways of presenting the answers. The message, however, is unmistakable. In broad terms, if all the cars in Western Europe were of today's designs, some 3 million tonnes of crude oil would be needed to produce their plastic components, but the annual consequential saving would amount to around 30 million tonnes.

The problem of traffic in the cities goes beyond considerations of fuel economy. It is abundantly clear that the automobile as we know it is a very unsuitable vehicle for use in crowded city streets. The search for viable city car designs has been proceeding for many years: the sticking point has always been the very poor performance/weight ratio of battery-carrying electric vehicles. Consequently the trend today is towards hybrids, using, for example, batteries for urban driving and conventional fuel for the country. The first volume production of this type of vehicle is likely to be the VW Chico, expected in 1995. Volvo are reported to be developing an environmental concept car with a gas turbine engine, which drives an electric generator linked to a motor which drives the wheels. It also has an aluminium body.

This choice of aluminium emphasizes the importance of weight saving. Plastic and composite bodies are less spoken of in such concept vehicles today than in recent years. This is primarily due to fears of the impracticability of recycling. Interest in the high volume use of plastic car bodies may not revive until the recycling issue is seen to be approaching a satisfactory conclusion. By then, however, the more serious issues of reducing fuel consumption and exhaust gases will have become more urgent, and will serve to emphasize the ecological benefits of plastics.

Environmental pollution

Emission regulations are building up, focusing on reducing carbon monoxide, oxides of nitrogen and unburnt hydrocarbons. (This is quite apart from the principle 'greenhouse gas', the carbon dioxide arising from normal petrol combustion.) Systems for Exhaust Gas Recycling (EGR) and Secondary Air Supply (SAS) are already commercial, and providing opportunities for plastics which can operate in the 140 to 200°C temperature range without loss of performance (see Chapter 6).

The weight saving argument for plastics embraces exhaust emission as well as fuel consumption, because the two are directly related, and the case for plastics is well proven. Nevertheless, this aspect hardly registers in popular awareness.

Another environmental trend with a bearing on plastics usage is the use of water-based paints. This movement is very strong, because of the desire to reduce solvent emission. Its effect on plastics is likely to be adverse, because topcoat oven temperatures are tending to become higher rather than lower, thus limiting the choice of plastics for car panels to certain more expensive materials.

Plastics and future design

In the immediate future we can expect evolutionary rather than revolutionary change. The automotive industry is essentially conservative anyway: it has to be, because of the enormous cost of change and the small margins in relation to the capital deployed. Continuing tight competition (and over-capacity) will lay stress on improved passenger comfort and process engineering, and maximum cost effectiveness. All these factors will encourage the greater use of plastics. Environmental awareness will continue to increase, though not necessarily accompanied by increased understanding! Probably, until the crisis in car shredding and fluff disposal is seen to be resolved, and some of the worries about incineration calmed, the environmental dimension will work against plastics.

Most forecasters agree that during this period the plastics likely to gain a larger share of the automotive market are: (1) PP, being cheap and relatively easy and 'clean' to recycle, (2) engineering plastics such as PA, PBT, PC and PPO, especially as blends for big volume applications like bumpers and body panels, (3) elastomers, and (4) GMT, based largely on PP. PVC, in spite of its versatility, will decline, in the face of concerted 'environmental' hostility. ABS and phenolics are already declining in their standard forms, but modified, application-targeted versions of both are staging a determined fight-back. PUR in its various manifestations should see increased usage as faster reaction rates are achieved, but many recycling aspects still need to be resolved, and the associations with CFCs will damage the growth prospects of PUR for some time. SMC has often been written off by some pundits, but it has a habit of surviving. New products and improved processes will ensure this.

Electronics applications will continue to grow: they confer obvious advantages in economy, handling and safety, and new road space management systems could solve some of the urban traffic problems. This is not an area which will use huge volumes of plastics, but there is likely to be useful market growth for polymers which offer high performance at high temperature, without excessively high prices (see Chapter 7).

In the longer term, in an effort to prevent the threatened 'gridlock', there will be pressure to ban private cars altogether from city centres. Vehicle designs for the 'open road' and the city centre will surely develop then along divergent paths. Hopefully, the eagerly awaited city cars will not come into the world loaded with the existing 'cultural luggage' of the motor car. Insistence on Class A finish, for instance, creates great difficulties for plastics, and adds very considerably to costs. Pitched well below the performance, trim and cost specifications of today's highly sophisticated 'small car', there should be room for simple one- and two-seater versions of the city car, with plastic bodies.

This a fertile ground for designers. In considering materials and production methods for this new class of vehicle, no possibility should be excluded. There are examples from the past, discredited perhaps for quite different reasons, from which lessons could be learned. The much-derided East German Trabant, as shown in Fig. 9.1, may have been the most polluting car ever, but its body pressings in resinated regenerated cellulose were cheap and functional. A similar production method based on pressed panels in modern materials could

9.1   Trabant.

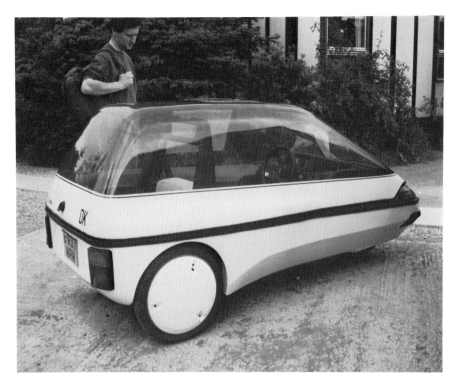

9.2   Mini-El.

9.3   Renault-Matra Zoom (courtesy GE Plastics).

be cost-effective at the right volume. The Sinclair C5, powered by a washing machine motor, has become a legendary marketing disaster, but it did have a well-designed and well-made body injection moulded in a PP copolymer.

Several prototype electric vehicles appeared during the 1980s. One of the more successful was the Danish 'Mini-El', with a body based on thermoformed acrylic sheet (see Fig. 9.2). The Renault-Matra Zoom of 1992 (see Fig. 9.3), exotic because of its adjustable wheelbase but nevertheless practicable, uses composite body panels.

Somewhere in the space between the two extremes of capacity and performance specifications is the traditional 'family car'. For most of the motoring public there is a sharp contrast between the projected fantasy image of the car held by its owner, and his or her actual lifestyle. The dominant impressions are of a high speed capability relevant only to the German autobahn, and of a showroom culture of opulence and high gloss. Media wisdom has encouraged unrealistic ambitions, whilst neglecting the ecological needs of the customer, and (almost invariably) debasing the benefits of plastics. It is true that the top-of-the-range cars provide the stimuli for new technology and new styling, as well as much of the necessary profit margin. It is also true that the real requirements for most people, which plastics do so much to deliver, are reliability, economy and durability, ensuring personal mobility. This, after all, is what it is all about.

Appendix 1: Polymer abbreviations and trade names

Abbreviation	Chemical name	Common name	Supplier	Trade name
ABS	Acrylonitrile butadiene styrene	—	GE Plastics Monsanto Dow Bayer Enichem DSM BASF	Cycolac Lustran Magnum Novodur Pavikral/Urtal Ronfalin Terluran
ASA	Acrylonitrile styrene acrylate	—	BASF	Luran
ACM	Acrylic acid ester rubber	—	BF Goodrich Cyanamid Du Pont Enimont	Hycar Cyanacryl Vamac Elaprim
BMI	Bismaleimide	—	Rhone Poulenc	Kinel
CA	Cellulose acetate	—	Courtaulds Eastman Kodak	Dexel Tenite
CAB	Cellulose acetate butyrate	—	Eastman Kodak	Tenite
EP	Epoxide	Epoxy	Ciba Geigy Dow Corning Scott Bader Shell	Araldite/Redux D E R Crystic Epikote
EPDM	Ethylene propylene diene	—	Bunawerke/Huls Enimont BF Goodrich DSM Du Pont Uniroyal	Buna Dulrat Epear Keltan Nordel Royalene
ETFE	Ethylene tetra fluorethylene	—	Du Pont Hoechst	Tefzel Hostaflon
FEP	Fluorethylene propylene	—	Du Pont Hoechst	Teflon FEP Hostaflon FEP

176

Appendix 1 *(Continued)*

Abbreviation	Chemical name	Common name	Supplier	Trade name
	Ionomer	—	Du Pont	Surlyn
PA	Polyamide	Nylon	Atochem	Rilsan (11)/ Orgamid (6)
			BASF	Ultramid (66,6,6/6T)
			Bayer	Durethan (66,6)
			BIP	Beetle/Jonylon (66,6)
			DSM	Stanyl (46)
			Du Pont	Zytel (66,612)/ Minlon (min.fd)
			EMS	Grilon (66,6)/ Grilamid (12)
			Enimont	Nivionplast (66,6)
			Hoechst	Hoechst/Celanese (66,6)
			Huls	Vestamid (12)/ Trogamid (6-3-T)
			ICI (now Du Pont)	Maranyl (66,6)/ Verton (long fibre)
			Rhone Poulenc	Technyl (66,6)
			Solvay	Ixef (MXD-6)
	PA based blends: PA + PE + PET PA + PPE (PPO)	—	Du Pont GE Plastics	Bexloy Noryl GTX
PAr	Polyarylate	—	Amoco Hoechst Solvay	Ardel Durel Arylef
PAI	Polyamide-imide	—	Torlon	Amoco
	Polyarylamide fibres: Aramid Meta-aramid		Du Pont Du Pont	Kevlar Nomex
PBT	Polybutylene terephthalate	—	BASF Bayer GE Plastics Hoechst Huls	Ultradur Pocan Valox Celanex Vestodur
	PBT based blends: PBT + ASA	—	BASF	Ultrablend
PC	Polycarbonate	—	Bayer Dow GE Plastics	Makrolon/Apec (polyestercarb.) Calibre Lexan

Appendix 1 *(Continued)*

Abbreviation	Chemical name	Common name	Supplier	Trade name
	PC based blends: PC + ABS	—	Bayer Dow GE Plastics	Bayblend Pulse Cycoloy
	PC + PBT		GE Plastics	Xenoy
	PC + ASA		BASF	Terblend
PE	Polyethylene:	—	Atochem BASF	Lacqtene Lupolen
HDPE	High density		BP Chemicals	Novex/Rigidex
LDPE	Low density		Dow	Dowlex
LLDPE	Linear low density		DSM Orchem Enichem Hoechst Orchem	Stamylan Loctrene Eraclear Hostalen Loctrene
PEEK	Polyether etherketone	—	ICI	Victrex
PEK	Polyetherketone	—	Hoechst BASF	Hostatec Ultrapek
PEI	Polyetherimide	—	GE Plastics	Ultem
PES	Polyethersulphone	—	BASF	Ultrason
PET	Polyethylene terephthalate	—	Akzo BIP Du Pont ICI	Arnite Beetle Rynite Melinar
PF	Phenol formaldehyde	Phenolic	BP Perstorp Rhone Poulenc	Cellobond Nestorite Resophene
PMMA	Polymethyl methacrylate	Acrylic	ICI BASF Rohm & Haas Du Pont Roehm	Diakon/Perspex (sheet) Lucryl Implex Lucite Plexiglas
PMMA-PUR		Acrylic thermoset	ICI	Modar
POM	Polyoxymethylene	Acetal	BASF Hoechst Du Pont	Ultraform Hostaform Delrin
PP	Polypropylene	—	Atochem BASF DSM Himont Hoechst	Lacqtene Novolen Stamylan Moplen Hostalen

Appendix 1 *(Continued)*

Abbreviation	Chemical name	Common name	Supplier	Trade name
PP *(continued)*			Huls	Vestolen
			ICI	Propathene, Procom
			Shell	Carlona
	PP foam	—	BASF	Neopolen
	PP glass mat	GMT-PP	BASF (Elastogran)	Elastopreg
			GE Plastics	Azdel
PPE (PPO)	Polyphenylene ether based products:			
	PPE (PPO) + PS or SB	—	BASF	Luranyl
			GE Plastics	Noryl
	PPE (PPO) + PA	—	GE Plastics	Noryl GTX
PPS	Polyphenylene sulphide	—	Bayer	Tedur
			GE Plastics	Supec
			Hoechst	Fortron
			Phillips	Ryton
			Solvay	Primef
PS	Polystyrene	—	Atochem	Lacqrene
			BASF	Polystyrol
			BP	BP Polystyrene
			Dow	Styron
PSU	Polysulphone	—	Amoco	Udel
			Ultrason	BASF
PTFE	Polytetrafluorethylene	—	Du Pont	Teflon
			Hoechst	Hostaflon
			ICI	Fluon
PUR	Polyurethane	—	Bayer	Desmopan/Bayflex
			BASF (Elastogran)	Elastoflex/Elastolit/ Elastofoam
			Dow	Specflex
			ICI	(ICI PUR)
			Shell	Caradol
PVC	Polyvinyl chloride	Vinyl	Atochem	Lacovyl
			BASF	Vinoflex
			EVC	Corvic/Vipla
			Hoechst	Hostalit
			Hydro	Hyvin
			Shell	Carina
SAN	Styrene acrylonitrile	—	BASF	Luran
			Dow	Tyril
			Monsanto	Lustran
SMA	Styrene maleic anhydride	—	DSM	Stapron

Appendix 1 *(Continued)*

Abbreviation	Chemical name	Common name	Supplier	Trade name
UF	Urea formaldehyde	—	BIP	Beetle/Scarab
UP	Unsaturated polyester	—	BIP BASF Cray Valley DSM Scott Bader	Beetle Palatal Synoject Flomat/Freeflo/ Stypol Crystic
	UP based SMC and BMC	—	BASF DSM Resins	Palapreg (DSM, SMC, etc)
VE	Vinyl ester	—	Dow BASF	Derakane Palatal
	VE based SMC and BMC	—	BASF	Palapreg

Sources: *Can it be made in plastics or rubber?*, Plastics and Rubber Advisory Service, London; H J Saechtling, *International plastics handbook*, Carl Hanser Verlag, Munich, 1987.

Appendix 2: PC based polymer databases

Name	Available from
PLASCAMS (Materials selector)	RAPRA Technology Ltd, Shawbury, Shrewsbury
CHEMRES (Chemical resistance selector)	RAPRA Technology Ltd, Shawbury, Shrewsbury
CAPS (Polymer selection)	Polydata, Dublin
CAMPUS (Material preselection by uniform standards)	Plastics raw material suppliers. Data are mutually compatible, and compatible with CAPS
MAT.DB (ASM International)	Comline Ltd
Adhesives only:	
PAL (Adhesives locator)	Permabond Adhesives Ltd, Eastleigh, Hants
EASel (Engineers' adhesives selector)	The Design Council, London

Sources: D Price, 'A guide to materials databases', *Materials World*, July 1993, p 418; The Materials Information Service, The Design Council, London.

Bibliography

Books

M Kaufman, *The first century of plastics*, The Plastics and Rubber Institute, London, 1963.

J Murphy, *Plastics and elastomers in automobiles*, Techline Industrial Data Service, London, 1993.

P C Powell, *Engineering with polymers*, Chapman & Hall, London, 1983.

H J Saechtling, *International plastics handbook*, Carl Hanser Verlag, Munich, 1987.

R Wood, *Automotive engineering plastics*, Pentech Press, London, 1991.

G Woods, *The ICI polyurethanes book*, John Wiley & Sons, Chichester, 1990.

Review papers and monographs

M G Bader, 'Polymers for advanced applications', University of Surrey course papers, 1993.

Ferruzzi Group Economic Research Dept, 'Chemistry and new materials – Montedison and the automotive sector', Ufficio Studi Ferruzzi, Milan, 1990.

C R Fussler, 'Ecobalances', Dow Plastics, 1990.

K V Gotham and M C Hough, 'The durability of high temperature thermoplastics', *Rapra*, 1984.

B Krummenacher, 'Automotive polymers and the environment', Dow Plastics, 1990.

B Krummenacher, 'Polymers in cars: energy consumption and recycling', Dow Plastics, 1990.

J Maxwell, 'Plastics in high temperature applications', *Rapra Review* 1992 **5** (8) no 56.

J D Robinson, 'Fundamental design for injection moulding', *Plastics* 1967 p446.

R C Stephenson, S Turner and M Whale, 'Flexural anisotropy and stiffness in thermoplastic sheet materials', *Plastics and Rubber* February 1980.

S Turner, 'Creep in thermoplastics', *British Plastics* June 1964.

G Walter, 'Status and prospects for plastics applications in automotive construction', *Kunststoffe* 1990 **80** (3) p293.

N A Waterman, 'Selecting the right plastic', *Plastics in Engineering* 1990 pp7–36.

'The major polymers explained', *Plastics in Engineering* 1989 pp12–61.

Conference papers

VDI (Verein Deutscher Ingenieure) Kunststofftechnik Conferences, Mannheim (VDI-Verlag, Dusseldorf):

K-D Johnke and P Behr, 'HDPE fuel tanks' (1982, p153).

E Hellriegel, 'Integrated front and rear bumper' (1983, p49).

D F Gentle, 'Manufacture of large thermoplastic panels' (1983, p145); 'Plastics in the engine compartment' (1983, p227).

G Walter, 'Possibilities and limitations of thermoset and thermoplastic usage in motor vehicles' (1986, p1).

J Tomforde, 'Car design and plastics: the conflict between reality and wishful thinking' (1986, p87).

J Adler, 'Tradition and innovation: Interior trim of VW cars' (1988, p77).

H Burst, U Tautenhahn and F R Wierschem, 'Plastics in the air bag system of the Porsche 944' (1988, p145).

H-G Haldenwanger, M Schneeweiss and M Maier, 'Development and testing of a fibre composite twin control arm rear axle' (1988, p251).

B Woite and G Schonleber, 'The BMW Z1' (1988, p275).

C Razim and C Kaniut, 'Innovation via modern materials engineering in automobile construction industry' (1989, p7).

J Kretschmer, 'High strength fibre composite components in the automotive industry' (1990, p177).

M J Wutz, 'Solutions for recycling plastic from scrap automobiles' (1990, p293).

'Plastics on the Road' Conferences, organised by PRI and I Mech E, London and Solihull
(Institute of Materials, London):

R J Murchie, 'Plastics under the bonnet; their future trials and tribulations' (1984).

V Nepote, 'Plastics in bodywork and trim' (1984).

A Weber, 'Prospects for large bodywork components and new possibilities for plastics in car interiors' (1984).

M F Saxby, 'Painting plastics' (1984).

A A Adams, 'Vehicle engineering for composites' (1986).

T E Creasey and J Maxwell, 'High performance composites for engineered components' (1986).

J F Yardley, 'Plastics in engines' (1986).

H R Orth, 'Acoustic properties of PP components in car interiors' (1988).

A S Wotherspoon, 'Automated injection moulding with lost metal core technology' (1988).

R Smethurst, 'Thermoplastic materials and processing techniques for large part high volume automotive production' (1988).

A Weber, 'Plastics in automotive engineering and aspects of plastic waste' (1990).

H Heyn, 'Blow moulding parts from engineering resins' (1990).

J Breitenbach, 'Bayer auto plastics seat development' (1990).

M Trueman and F Neutens, 'Under the bonnet applications of engineering thermosets' (1990).

U Lehmann-Burgel, 'Vinyl ester resins for high performance automotive applications' (1990).

S Lepore, 'Research and development of composite materials in Fiat' (1990).

B Cummings, 'Fibre reinforced plastic composite engine' (1990).

G F Smith and T G Parr, 'Design and production of a plastic inlet manifold' (1992).

M R Hartland and H E Rowlands, 'Development of the roof structure for the Vauxhall Frontera Sport' (1992).

H J Schutt, H Junker and A Seewald, 'Development and application of FRP suspension components' (1992).

E Renfordt-Sasse, 'Blow moulding of external components for motor vehicles' (1992).

D F Gentle, 'End of life vehicle disposal; a car industry viewpoint' (1992).

P Orth, B Keller and A Schmiemann, 'Material suppliers' viewpoint on the recycling of engineering plastics' (1992).

J R Whittaker, 'Viewpoint of the recycler' (1992).

Auto Tech Congresses, Birmingham
(Institute of Mechanical Engineers, London):

E M Rowbotham, 'Plastics under the hood' (1985).

D I Wimpenny, 'Characterisation of thermoplastics for practical design' (1991).

M A Potter, 'Structural RIM' (1991).

A A Bennett, 'Recycling and auto engineering for the new millenium' (1991).

J R Whittaker, 'Vehicle recycling – the car shredder's perspective' (1991).

A J L Hutton, 'Energy balance – automotive materials choices' (1991).

ATA (Associazione Technica dell'Automobile) Seminars, Turin:

J Cusset, 'Development of semi-structural and multifunctional plastic body parts within the Peugeot SA Group' (1987).

P Beardmore, 'FRP composites for future automotive construction' (1987).

B Krummenacher, 'Production of composite body panels and structural parts by RTM' (1987).

J F Monk, 'Why polyester DMC is replacing steel for headlamp reflectors' (1989).

F Cassese, A Garro and G B Razelli, 'State-of-the-art construction materials for high performance cars' (1989).

J Maxwell, 'Thermoplastic composites for high performance automotive applications' (1989).

H Muller, 'PUR/glass mat composites for automotive applications' (1989).

W I Spurr, 'A low smoke, low toxicity fire retardant system for public transport applications' (1989).

SAE (Society of Automobile Engineers) Annual Conference, Detroit, 1987
(SAE Inc Publications Division, Detroit):

M Kallaur, 'Composition of automobile body panel materials'.

D G Dumouchelle and R A Florence, 'Advanced instrument panel materials and process'.

R Weibner and J Adler, 'Plastics for interior trim of passenger cars, present and future'.

R Eller, 'Materials substitution in automotive interiors'.

Y Kurihara, K Nakazawa, K Ohashi, S Momoo and K Numazali, 'Development of multi-layer plastics fuel tanks for Nissan research vehicle'.

V Nepote and F Rossi, 'Ten years of applications in Fiat car exterior components'.

SIA (La Societe des Ingenieurs de l'Automobile) Conferences, France:

  B Bertrand, 'The automobile engine towards the next century – which materials?'
    (1988).
  G Buisson, 'The European market for composites in car bodies' (1990).
  M Marcellas, 'Resin transfer moulding technology in Matra Automobile' (1990).
  S Glommeau, 'Evolution of exterior plastic components on Renault vehicles'
    (1990).
  J-P Bauchel, 'Plastic painting in the automotive industry' (1990).

'Automobile, Chemistry, Design' Symposium, 1989
(Societe Industrielle de Mulhouse):

  U Bahnsen, 'Evolution of vehicle design'.
  P Gaborit, 'Synthetic materials for large body parts'.
  G Secheresse, 'Concepts in thermoplastics'.
  H N van den Berg, 'Contribution of engineering polymers to styling evolution'.
  H Grune, 'PUR for automotive skin panels'.
  J-L Caussin, 'From "Plastoc" [*plastic "tat"*] to composites'.
  G Kranz, 'Engineering plastics for automotive headlamps'.
  D Prevot, 'Breakthrough in structural thermoplastic composites (STC)'.
  C R Fussler, 'Polymers for body panels'.

Chemistry and the Automobile Conference, Barcelona, 1989
(Congreso de Quimica del Automovil):

  P Salomon and R Wilkens, 'Innovative use of plastics for seats'.
  F Gubitz, 'New polymers for the automotive industry'.
  A Weber, 'Technical aspects of applications and recycling'.

PRI 'Plastics Recycling – Future Challenges' Conference, London, 1989
(Institute of Materials, London):

  E M Rowbotham, 'The motor car and recycling'.
  S S Labana, 'Use and recycling of plastics in the automotive industry in the USA'.

Technical and promotional literature

Technical and promotional literature was obtained from the following plastics
  manufacturers:

  Amoco, BASF, Bayer, BIP, Dow Plastics, DSM, Du Pont, Elf-Ato, Ems, Exxon,
    General Electric Plastics, Himont, Hoechst, Huls, ICI, Owens Corning, Philips,
    Rhone Poulenc, Solvay.

Index